A HOT TEA
BY THE GIZA

THE REAL GLOBAL
WARMING, NOT CO2 HOAX

Nae Ismail

iUniverse, Inc.
New York Bloomington

A HOT TEA BY THE GIZA
THE REAL GLOBAL WARMING, NOT CO2 HOAX

iUniverse books may be ordered through booksellers or by contacting:

iUniverse
1663 Liberty Drive
Bloomington, IN 47403
www.iuniverse.com
1-800-Authors (1-800-288-4677)

ISBN: 978-1-4502-3819-9 (pbk)
ISBN: 978-1-4502-3820-5 (ebk)

Printed in the United States of America

iUniverse rev. date: 7/15/10

TABLE OF CONTENTS

FIGURES

Quotations

There is no seniority in learning.

Whoever attains understanding is the teacher.

Chinese Proverb.

Life is a journey of making choices

Some are cerebral, others pure gut instincts

Destination depends on what ratio between them one chose

Nae Ismail.

Dedication

To my parents Imam Mohammed Rashid Ismail (Ma Je Chiu), Salimah (Shiu Hing) Yeung of Canton (Guangzhou), Macao and Hong Kong plus all my siblings for their love, patience, advice, support and sacrifices made in war and peace times to ensure that I could obtain a university education as the first in the family. I owe them each and every word that I can write and have written.

Preface

Nothing epitomizes six decades of continuous learning like condensing them into a book. There is fun in learning, and having fun is natural for the inner child in us. Learning of course has a serious side too. This is my effort to share with you what I have learnt in a few topics below, some of which may have stumped you as they did me for a while.

- CO_2 threat to global warming and Gaia* Theory
- Earth's magnetic field and its relationship to the Ionosphere.
- how a tornado or waterspout or hurricane is spawned
- Coriolis twist, Bermuda Triangle Mystery, and how Amelia Earhart vanished
- radiation physics and plasma physics
- what do Jet Streams and Mars ferocious dust storms have in common
- energy conservation law amendment and why perpetual machines flop
- a scientific definition for Energy and how electricity actually flows

*Note. Name of a mythological Greek goddess, Gaia theory suggests that all life forms share a spiritual connection and influence each other physically. Earth is regarded a life form (Big Mama) that protects her young.

Of our five senses, sight is the busiest and most trusted. Eyes work involuntarily in each waking moment. Located below the frontal lobe of the cerebral cortex for least process time, eyes detect either threat or opportunity farther and earlier than other senses to give us the best chance to prepare or react, often in life and death situations. It is no coincidence that sight ranks top for survival. Seeing-is-believing has become such truism therein lurks the risk of unintended self-deception. Our low visual resolution limit convinces us to see an object that in fact is hundreds of thousands times emptier than it is solid. Shadows cast by light lead us to conclude that sunrays are photon beams projected in straight lines. It is my intention to point out such fallacy to you so that you will see a real world differently after reading this book.

Originally titled "The Energy Myth" but upon reflecting how I was inspired into writing it, I decided on a 1965 incident as a more fitting heading for the book. Backpacking then in Cairo I stumbled upon a Bedouin camel ride guide by the 4000 years old Great Giza pyramid. He served me a very hot tea tossed back and forth between two tin cups and in his action taught me unknowingly then how our planets stayed cool throughout history despite 24/7 solar bombardment. It did not dawn on me until over four decades later.

This book is about my understanding of *energy*, not weather. Weather is about air movement (wind) and effects of moisture in air. Weather is used here in common interest to illustrate how solar energy and gravity combine for its creation. There would be no weather without sunshine *and* gravity. Old theory attributed 100% of weather to the sun, which should mean a calm quiet winter when it is farthest away. That is so untrue. In reality, winter can feature worse blizzards than summer tornadoes. Where is the energy from? Earth gravity is responsible as you shall see. Neither a climatologist nor meteorologist by training, I have taken generous liberty of their tolerance and good grace to delve into climate discussions as seen through my energy lens. Where I erred in their professional eyes I hope that they cut me enough slack with a forgiving grin, or a nod. We want global warming* reversed for good reasons, but mainly to give future generations their needed knowledge and effective tools to deal with it after we are gone too early to help.

*Note. Warming does *not* have to be literally a temperature rise. It refers to an overall energy content increase in the system. There is a difference. Temperature is adjustable by humidity like weight depends on gravity. On the moon, we weigh only one-sixth as we do on Earth by its less gravity. Using temperature to gauge energy content without knowledge of humidity along side is scientifically naïve. It is as pointless as clocking a car road speed to guess its engine RPM (revolutions per minute) if neither its tire size nor transmission ratio at any moment is known. Energy can bring cold just as easily as warmth. Isn't that amazing ?

More than just belief, I now understand *how* Earth climate is changing but reject (CO2) carbon dioxide emission theory as its cause. Proof of human involvement in warming is irrefutable and overwhelming. Climate change deniers are closing their eyes to what is real like cigarette smokers in the past scoffed at lung cancer warning of doctors, and irresponsibly ignored the harmful impact of their habit to the innocent mass. Climate debate has degenerated into semblance of NHL hockey fracas and UK soccer hooligan melees. It is embarrassing and fruitless. We must restrict it back to a scientific forum without raucous media, politicos, economists, pundits and lay(wo)men in the fray. Keep them out. Scientists must be the only jurists as they ought to be. Granted they have done a lousy job so far. Lock them up in the laboratory and keep the key until they find the right answer.

The book explains facts as I know them and extended by my personal insights through in-depth reasoning. Readers are encouraged to research for facts for your own verification and not take my ideas for granted. I took liberty in advancing **original** concepts of mine neither as fantasy nor lay imagination but as scientifically grounded defendable extensions of facts in order to explain weather observations that are currently unexplained in run-of-the-mill literature. Should I turn out to be correct, well and good that I have moved science ahead. I am ready to accept sensible and polite challenge to their accuracy from critics but only if they furnish demonstrable proof to the contrary. Science is a civil dialogue and cooperation of minds. Who is right should not be any issue. Truth stands on its own merits with debates or not.

My inclination towards science surfaced early. Barely age five, I stumbled upon a detail sketch of a jet plane structure in a magazine. Without prompting I meticulously copied it enlarged an hour later on a chalkboard that my dad used in teaching Quran class to Friday Juma congregations. Six decades since the sketch is still as vivid in my mind as was then. That episode never inspired me to be a pilot even after having logged enough air miles since then to circumnavigate the global at least four times. It did open floodgates to my imagination and sense of wonderment that took me higher, deeper and farther than any magic carpet could have physically delivered me. The mind is our best travel companion. It is timeless but cumulative, cheaper but enriching, unrestrictive but constrains us to choice by personal criteria, unhindered by terrain but expansive over any field of interest. Give it free rein with just one strict proviso to abide by logic or gut instincts if a junction looms up as if consulting a road map on a real trip. Its thrill can sustain you an entire lifetime.

From the start, I had serious misgivings about attempting this book. Advancing age and health issues raised red flags if it could be finished. I have distaste for incomplete works that reflect non-commitment. Yet, I also felt duty bound to sound a clarion call witnessing current global warming struggle as wrong headed that will not bear fruit. Caught between a rock and the sea, I resolved at last to the same decision I took in spring of 2002 after surviving an ischemic stroke. Learning to walk again, I focused on "3-steps ahead" at a time, and left the rest to Divinity. I wrote little by little what I could. I edited where it warranted. A notion floated up during writing that if nothing else it would be when finished a private one-sided conversion to our grandchildren and to theirs in the future. I wrote it purposely at a level suitable to the comprehension ability of perhaps a 16-year old adolescent. For my part as a university post-graduate physicist, I have fulfilled an obligation to give my best insight on the climate change issue and removed some stumbling blocks hopefully. Researching for the book, I sensed a possible holistic connection* between weather above ground perhaps intricately inked to tectonic plate activities like volcanic eruptions and earthquakes. That led to my realization of Earth's magnetic field with a new origin.

*Note. Details in Chapter 7 (Earth gravity, Coriolis Twist, Magnetic field, etc)

Life journey can run parallel to a road trip. Peregrinating from point to point we round corners, switch lanes, speed up, slow down, apply brakes, dodge obstructions and *make decisions* each step along the way. At times, the scenery takes our breadth away on a smooth and serene ride. Other times we splash through potholes, bumps, mud, dusts and foul weather, cursing that we must have taken a wrong detour somewhere. Surprises and unpredictability make a journey refreshing, sometimes even more than final destination. Surprises register while the planned slips by ignored. Who would have guessed that a social hot tea served next to a 4000 year old pyramid was the main cue in explaining half a century later how planets have stayed cool without burning up in the solar system? I feel so very privileged.

Curiosity has been a driving force all my life. Asking *myself* questions how or why, I am not easily satisfied until everything "hangs together" properly without contradictions. This tenacity for truth has helped me sharpen a critical analytical skill in deciding what makes sense, or not. Summer 2007 was one of those turning points in my life. I decided to re-read a few university physics texts in a review of what I had been taught. To my surprise, I found myself disagreeing with regurgitated aphorisms based on what I had learnt later in practice. That primed me with the motive of verbalizing my personal insight in a book to challenge the status quo. I have opted to bypass rigorous mathematical treatment of topics included in the book, reconciling to myself that it is best left as theses material for prospective graduate students. I want to neither dampen readers' interest nor slow down their reading by tedious formulae or equations. My aim is a large general audience who is equipped with high school science (science teachers in particular who are charged to lead their flocks on a path built on veritable facts and not blind faiths) and seeks for a better understanding of events in their daily lives and the physics behind their occurrence, yet unwilling enough to sweat through boring calculations. I tried my best to write in plain language instead of highfalutin technical mumbo-jumbos so any future grandchildren can read what I am telling them.

What is most at stake is for our bright students to align their thoughts and logic properly to reach right conclusions that they would accept voluntarily instead of gulping down undigested dicta dished out by intimidating professors. CO_2 greenhouse gas (GHG) controversy is a sad example of how false science can elude scrutiny, as it did, spread like a virus and be regurgitated for centuries before it is exposed at last after much damage done and precious time wasted. Worst of it is polluting and confusing precious young minds. Ironclad proof of a new science should be in place **before** its technology can be exploited ethically. It may be a tall order and hard to put into practice when economics decides livelihood. Flying blind is never a smart move. We have done that too many times so far and luckily escaped unscathed. Our luck may not hold out this time.

In the end, this book is a journey of self-discovery*. Everyday, folks climb Mt. Everest, trek over Sahara, fly balloons over the Alps, swim across

English Channel or five Great Lakes to test their mettle. Having lost my mobility to a stroke in spring of 2002, I just did it differently with not my limbs but my brain, the very organ that crippled my limbs. What an irony! The trials and tribulations are no less trying in the struggle, but worth every groan of pain and frustration that was endured. I relished it without regrets. Above all, when our grandchildren read it one day I hope that they will appreciate what I am leaving to them.

*Note. Football has dead zones at both ends. Once the ball goes beyond goal posts into end zones, all bets are off and field game rules no longer apply. If an asteroid smacks into a planet, it is end zone for asteroid but the planet carries on as if nothing happened. When a bullet finds its living mark, it is end of story for both. Any death is an end zone. I learnt much to my chagrin that dead zone also applies to conservation laws and rules in physics. That jolted the idealist scientist in me. Each time the hot tea hit the cup in the Giza Bedouin's hand, more of its heat went into dead zone never to return. When energy comes to a stop, it is all over. Zip.

In my Islamic faith, a pre-ordained destiny is in our beliefs. We do not sit idly by in wait for events to unfold, but we are supposed to reserve private moments daily or nightly to introspect and tune to our instincts for directions to follow. Had I not picked a Canadian university for education, I would not have experienced freezing winters, touched white reflective snow and learned of black ice that is the curse of winter motoring. Had I not run into the Giza Bedouin camel guide/ tea maker, I would not have witnessed first hand how a hot tea cooled fast by heat converted to kinetic energy. Had I not in retirement regrouped with former colleagues Bill and Graham on a hunch to explore wind turbine technology, I would not have revisited Bernoulli cooling principle. Without doing solar PV research for a corporate employer initially and later in my private business, my solar knowledge would be limited to just lay. Finally without the stroke incident that forced me into retirement with time on hand and restricted me to cerebral thoughts in order to ignore my awkward gait during daily outside walks, I would not have devoted as much attention to questions and answers on climate physics. Should this book have any impact on energy science in some small way, I was possibly tasked by destiny to bring it forth all along. If for no better reason than to witness first hand the supreme intelligence behind a balanced design and craftily engineered Earth in which to live, I would gladly do it over all again any time. The possibility of randomness in making it happen is infinitesimal. Life is fragile. A tiny temperature slip of mere degrees off our 36.7 C body temperature could prove fatal. Moisture cushions us by moderating weather to enable human habitation. Without its mitigation, there is just no possibility for life on Earth to survive the extreme violence of a heat exchange. Dreams of colonizing Mars one day will likely remain fantasy. Its first dust storm would make Antarctica explorations like a summer picnic. I earnestly invite readers to draw your own conclusion how life came to be. Granted without water, life might have evolved differently but so far that argument has found no support on other planets. To me anyway, water is our Creator's signature of a higher supreme intelligence. To assume that homosapien possesses enough intelligence to fully understand the universe is nothing short of the ultimate hubris.

Evolution theory believes in a natural selection process of species to survive. In that argument, shouldn't there be life species evolved on any planet to adapt to a CO_2 or SO_2 or any kind of atmosphere ?

Dear readers;

World population was 3 billions after World War 2. It soon doubled to 6 billions at the start of 21st century. By 2050, it will triple to 9 billions by UN projection. Somewhere down the road, Nature may cull by a selection process. Only prescience and preparation can tip the outcome of "who goes or stays". We found energy, the Genie. Energy is our lifeline to modernity. What price are we willing to pay for the Genie as our slave? We won't know without accurate full knowledge of energy truth to replace current hearsay, rhetoric, distrust, guesswork, skepticism and suspicion. Energy users must act responsibly knowing full predictable outcome of their actions without excuses. The Copenhagen 2009 debacle (so called Hopenhagen or Brokenhagen) is cold evidence that after four decades we are still flying blind in the dark without new insight and arguing in circles over the same tedious question endlessly. Are we passing the morass over to our kids while basking in the sun ?

If you agree with the material presented in this book, that they have enlightened your comprehension in energy physics and how climate obeys laws of physics, please send your feedback, comment or suggestions for improvement to nm50909@rogers.com and donate to my website www.ahotteabythegiza.ca. in order to join and amplify my voice. Recommend the book to those who would benefit from having it. Encourage friends and family to peruse it. Book sales proceeds and donations will be directed to my fervent objective of assembling a group of like minded scientists to study "<u>Upper Atmosphere Low Density Solar Heated Plasma</u>" as a potential energy resource.

This timeless 24/7 solar power is potentially 5 times more powerful than current PV solar technology. I am optimistic that it can be tapped safely with creativity as the ultimate energy for humanity.

Time is of the essence. It is progressive thinking to replace status quo sponging off remnant fossil fuel like prodigal offspring that we are now.

This energy has enough intensity to wreak havoc like a tornado at its bare minimum. Unopposed, it can trigger earthquakes and eruptions. Taming and harnessing it will be a paramount achievement of epic proportions that rises above and beyond a safe journey to Mars and return of this century. It is entirely our call.

Our elaborate institutions are pumping out technologists to meet industrial needs, but at the expense of scientists who can help us better understand before it is too late where we are hurling head first like a MAGLEV (magnetic levitation) bullet train but minus a conductor. Your support is needed and will be appreciated by your children and theirs when they grow up realizing the predicament to which we have committed them with neither their consent nor even awareness. Thank you.

Chapter 1

The Scenario

Due diligence to verify facts and figures is the foundation of good business practice. It works. Good faith alone is an insufficient criterion. Deals should not be consummated without it in the process somewhere, at the start preferably to avoid wasting time and energy. This cardinal rule was broken, however intentional, by scientists (imagine of all disciplines) through negligence and sloppiness for the past two centuries that bewitched us with the longest quagmire foisted on mankind. The egg on our face is well deserved. The public has put faith on our ability, honesty, integrity, professionalism and above all reliability with incontrovertible evidence for truth. I am referring to the raging controversy on greenhouse gases (GHG) as a heat barricade that traps solar heat to bring about one day an infernal Armageddon, the carbon footprint hue and cry all over the world.

While 21st century global community is consumed with this threat, yet we are oblivious about a far more ominous thermal menace under our feet that could either instantly vaporize us in a jarring volcanic eruption or bit-by-bit roast us into pemmican like Jamaican jerky on rotisserie if not for an ingenious cooling design that has held it in check like Genie in a corked bottle for 4.6 billion years and kept a rock stable comfortable 15 C (=59 F) ground temperature in order to sustain life. Understanding this system reveals how Earth's ecosystem really functions and manages her energy. It also sheds light on a way out of the CO2 hole in which we have dug ourselves so ignorantly.

This book is not a spellbinding who-done-it that, once begun, is irresistible to put down. It is possible if you are technically conversant enough with all topics presented to follow the thread of my thoughts. For the majority of readers I recommend treating it as going to an all-you-can-eat buffet without time constraint. Read it in small doses but digest well and not hop all over its pages. Research other references on anything unclear to you. The only award at the end is an in-depth comprehension. There is no prize in finishing a good meal first. If it took me six decades to put it into words finally, it is not unkind as a tribute to me to finish in six days, six weeks, or even six months. Like a guided tour, I will escort you through a plethora of integrated parts in physics from solar radiation to plasma like pieces of a jigsaw puzzle. Once assembled, a picture would emerge that shows you how Earth climate works under energy laws. Professor James Lovelock in UK proposed a Gaia (ancient goddess) theory purporting that Earth is like a living creature. To the extent that all creatures live by energy, I am with him. Energy is life. Without energy, it is just a corpse. Where he and I part ways is that energy obeys physical laws without compromise and is not subject to wishes of anyone as a master.

We live in a very delicate environment like fleas on a dog. Should it twitch, it is earthquakes to us. Energy is all about action and therefore some degree of violence is inherent with energy. It cannot be helped. The essence is in control of that violence to be tolerable. Waste is a part of life cycle and energy disposal is part of the waste stream to permit fresh energy to repeat the cycle. If more waste is generated, more energy is accompanied in its disposal. This should put energy use into proper perspective when it comes to our fossil fuel consumption habits.

1965 was a memorable year for university students like me. The hottest selling book was "Europe on $5 a day". It detailed locations in city maps where low budget, $1 - 2/night, youth hostels sex segregated dormitory accommodations could be found on foot or by public transit. It was definitely doable. Guidebook in hand, backpackers by the thousands from N. America, Australia, and New Zealand who were hungry for discovery and eager to explore invaded Europe every summer with regularity like annual migration by the millions of monarch butterflies in October-November south to Mexico for winter hibernation. I did my share of roaming up and down Western Europe for 4 weeks. After Rome, I stopped at Cairo in transit to the Far East to visit family back home after 5 years of being away. I stayed in an inexpensive inn for Egyptians to save money.

Resident Muslims in the inn kindly taught me local bus routes to see the mind boggling Cairo Museum, the magnificent Citadel mosque, a coffee-and-cake hilltop vista where I met an army colonel as a new acquaintance, the endless labyrinth street stalls in the hustle-bustle Cairo Market square, and to Giza desert to pay homage to the Great pyramids and Sphinx. I made good friends and was having a wonderful time, often on their generous hospitality no less. I became a frequent Cairo bus passenger. On each bus ride, I marveled about the hanger-on free riders outside for their skill, precision and confidence. It was impressive. I remember itching then to try once just for the experience but chickened out for fear of injury that would ruin my travel plan. It was thrill enough seeing it nonetheless.

As any visitor to Egypt would testify, there was little to see inside a pyramid but empty chambers and low tunnels to crawl in from room to room. All artifacts had been removed to museums for safer keeping. Its interior did provide temporary relief from the choking heat outside. Most remarkable was the huge stones that made up the pyramid. Emerging from the pyramid into a blinding sun, I came upon a Bedouin camel ride guide soliciting customers and peddling postcards. Hearing my "Asalam Alaikum" salutation to him, he replied "Wa-alaikum Salam" and asked for my origin. I told him. He then offered me a "discount" on his camel ride plus invitation to have a chat and a social tea afterwards, which I accepted. After the camel ride and photography, the Bedouin ushered me to a spot in the shade of the pyramid where he had set up camp. He proceeded to make tea in a tin kettle on a small charcoal stove. Unlike anywhere else, Arabs boil tea leaves in the water to make it strong. When it came to a boil he used two tin cups to pour tea back and forth several times before handing a cup to me. Granulated sugar was added. Arabs love sweet sugary tea. I sampled the tea. It was strong, sweet hot but not scalding that I was prepared for. In mere seconds he had cooled it faster than any fan could have done. Heat is gone. This trick of nature is what drives our world and a subtle fundamental in energy physics was revealed to me as well.

That encounter was very pleasant but I quickly filed the episode to my brain archive to make room for other pressing events at the time. It laid dormant for 45 years until this year where suddenly it surfaced subconsciously on its own while I struggled to figure out how planets have kept cool despite nonstop solar radiation for eons. It dawned on me that the Bedouin tea toss between two cups was a demonstration of heat converted to kinetic energy with help of gravity. On impact, kinetic energy is dissipated and heat is gone. The hot tea behind the Giza pyramid was a lesson of Earth physics in display of ordinary life. I just did not grasp it at the time until many years later.

In catastrophes and devastation power, a hurricane (typhoon in Asia) can possibly run equal to earthquakes, volcano eruptions, landslides, floods and tsunamis. Each has comparable energy tonnage of the atom bombs that obliterated Nagasaki and Hiroshima of Japan in 1945. Destructive power is 100% **instant** kinetic energy resident in wind and in the storm surge. Its source though is 100% thermal in the moisture evaporated by sunlight off the ocean surface. Meteorologists have struggled to find exact explanation how hurricanes and tornadoes too are spawned for that matter. I will answer that in Chapter 9 later.

At the time this manuscript was submitted for review, news channels reported an 8.8 Richter Scale earthquake (February 27, 2010) just off the coast near Santiago, Chile while two separate blizzards lambasted both sides of north Atlantic coasts. A storm watch was issued as well for more snow and heavy wind over the Texas Gold Coast. The earthquake sent high waves 6,000 miles away to Hawaii. This occurred scarcely a month after a 7.8 Richter Scale earthquake in Haiti that leveled one million homes and killed an estimated 200,000 in the 9 million nation. Are these coincidences or something else is going on ? Read on if you desire answers that I will provide with scientific arguments and reasons.

Chapter 2

Matter and Non-matter

Matter

Matter appeals to our sight and sense of touch (pressure and texture) by its physicality. Occasionally a matter may even offer odor to signify its presence. Scientifically, matter has mass (not weight), size, shape, colour, hardness and other noticeable attributes that identify the object. Matter obeys Newton's First Law of Motion that says an object stays at rest unless it is under a force to move in some direction (action). Any action creates a reaction, which is a force in the reverse direction. Matter is comprised of particles that are called atoms (in a single element) and molecules (mixture of different atoms). I use a term "particle physics" to describe matter behaviour. Particle physics is purely force governed by Newton's Law. Particle motion trajectory is a manifestation of the kinetic energy embedded within. In wind, air molecules are invisible but their collective action is unmistakable. In waves, there is no question about either aspect.

At a set of temperature and air pressure, matter can exist in 1 of 4 possible states, namely solid, liquid, gas or plasma. Broken down to its smallest entity, matter consists of atoms or molecules. An atom or molecule has a nucleus made of neutrons (no electric charge) and protons (each has + ve charge) together at its core. Negatively – ve charged electrons whirl around the + ve nucleus in layers of distant orbits far away. Electron orbital layers have rising energy levels towards the nucleus. An electron closest to the nucleus requires X-rays (10 – 100 KeV) to be evicted while one at

8

the periphery would do by visible light (1 – 10 eV). Visualize electrons as cars and their orbits as marked *speed* lanes in which they travel, but never in between. A lane change for a car involves either acceleration or deceleration. Likewise, an electron *excited* by extra energy from a light photon (photoelectric effect) or collision by other electrons jumps to a higher orbit. The reverse is also true. A *relaxed* electron falls down to a lower orbit by giving up energy as radiation typically. If an excited electron leaves the molecule, ionization has occurred. The molecule minus an electron is now a positive charged + ve *ion*. In fluids like molten lava, ions are mobile by its fluidity. In solid, an ionized molecule (or atom) that stays in a fixed stationary spot is a *space charge* (not ion). A cluster of negative – ve charge electrons is called an electron "cloud".

Solid

As a solid, I think of it as a coconut in its husk. Forget the juice (whey) inside for now. Coconut has a *hard* edible white fleshy interior surrounded by a *loose* fibrous coir outside inedible husk. In metals, the coconut outer fibrous coir husk would be a flexible pliant electron web (with long and wide orbits) that shrouds the metal in a skin and permeates between atoms in a honeycomb lattice. The edible fleshy white part of coconut represents a rigid framework* of nuclei bonded to one another by intermolecular magnetic forces at close range. If heat is added to raise temperature, all nuclei get agitated by extra energy to shake violently. Shaking breaks the inter-nuclear bonds until the framework lets go and atoms move free independently. This is the Melting Point when a metal starts to flow and can split apart. Only the pliant electron web (the coconut fibrous husk) holds the atoms in place. As liquid, electron web integrity represents its surface tension to stop atoms from breaking free as gas. Continue heating to raise temperature higher, nuclei agitation gets strong enough to disintegrate electron web to let atoms go free as a gas. This is the Boiling Point. Heat of vaporization is a measure of liquid surface tension and heat of fusion indicates bond strength of inter-nuclear magnetic coupling between atoms.

*Note. Kids playing with magnets know how hard it is to separate two magnets with opposite poles (N-S) together, or conversely how hard to squeeze two like poles (N-N or S-S) against one another. In magnetism, coupling force rises up exponentially if the gap gets smaller. Heat destroys magnetism. At the Curie Temperature all magnetism is lost. Curie temperature is often just below Melting Point for this reason.

Magnetic nature of the inter-nuclear binding force suggests to me a nuclear structure that disagrees with conventional model. So far the nucleus is treated as an aggregate of neutrons (no charge) blended with protons (+ ve charge) together as a core. There is no satisfactory explanation as to how normally mutually repelling + ve charge protons would tolerate close proximity in the nucleus. I propose an alternate model that makes more sense to me (*Figure 1*). The Periodic Table of elements shows a 1 to 1.6

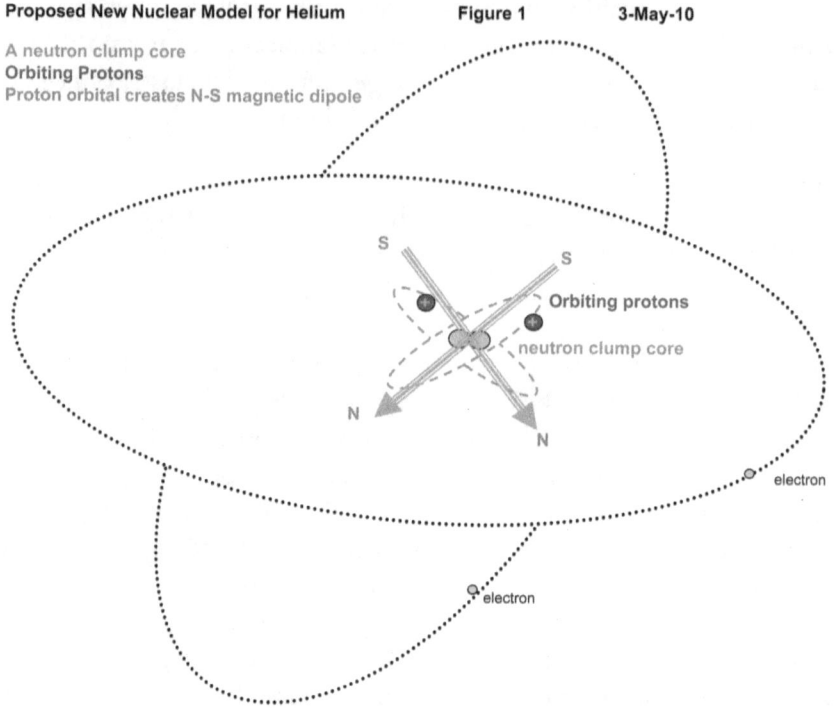

Figure 1. Excel graphic of Proposed Nuclear Model for Helium

neutron-to-proton ratio consistently in the nucleus, rising with atomic weight. There is at least equal or more neutrons than protons in a nucleus without exception. Suppose the no-charge neutrons clump tightly by gravity at its true core with protons orbiting around tightly in separate orbits. Structurally proton orbits would be a contracted version at the core but similar to electrons orbits further out at the atom edge. Orbiting protons would generate magnetic dipoles for inter-nuclear bonds and solve the dilemma of mutual repulsion of protons that are now moving extremely fast in tiny orbits. By the (1 – 1.6) mass ratio, orbiting protons would wobble neutron core into its own frequency of motion for energy coupling. Protons also obey quantized energy rules like electrons to be able to accept energy exchange with gamma rays (γ) at frequencies 10*6 times higher than visible light that are known to originate from the nucleus. Protons orbiting in tiny orbits around a neutron clump core must whirl at extremely high velocity to build enough centrifugal force to counter the gravity pull from the neutron clump core. There is good agreement from this argument.

Proton mass (M) is 1840 times that of electron. Its extreme orbital velocity (V) to create enough centrifugal force to oppose gravity pull by the neutron clump core means that a Gargantuan size of kinetic energy (½MV*2) is stored in its motion. This huge amount of energy manifests itself by release of extremely powerful gamma rays (γ) of 10 KeV to 10 MeV when a proton <u>decelerates</u> to make a daughter nucleus in a radioactive decay or fission of an atom such as 92 Uranium 238. A proton deceleration emits gamma rays as surplus energy by daughter neutron clump that is smaller with less gravity pull. A weaker opposing centrifugal force is sufficient from a slower V.

When two almost equal frequencies are close together, they beat and generate a new frequency of their difference. This beat frequency is added and subtracted to both parent frequencies as what are called their sidebands. Sidebands have energy. Two frequencies starting off become four (six in fact, but two are repeats) frequencies close together. Add more frequencies from nearby sources, the energy spectrum quickly blur into a solid band and not isolated frequencies any more. If an atom sends out one frequency, a solid body of atoms would produce a continuous band of frequencies capable of resonant coupling to radiations. That makes metals good energy absorbers and conductors because of their wide energy bands.

Conductor

Metals are good conductors of both heat and electricity, but not light. Conduction needs mobile charge (+ or -) carriers. In metals, conduction occurs by a − ve charge electron web. Conduction by an integrated but pliant honeycomb electron web throughout the material can move an energy *field* at the speed of light, c. Electrons can either interact with visible light as photons (Einstein's photoelectric effect) by resonance or blocking it (opaqueness). If painted black to raise their coupling factor (closer impedance match), metals can absorb light as heat.

Insulator

Electron web of an insulator has discontinuities, weak boundaries or junctions that offer high resistance to energy field flow and would rupture easily when stressed. That would occur in a hammer blow like a lightning bolt. Air is a gas mixture of nitrogen (78%) and oxygen (21%) with traces of carbon dioxide (0.03%) and others. Air is a well known insulator for both heat and electricity. Once ionized however (called plasma), its properties flip 180 degrees because *ions* are excellent *electrical and heat* absorbers and conductors. This fact is shown later in how sunlight generates strong climates in the upper atmosphere.

Liquid

A solid retains its shape and form while both liquid and gas do not. The latter two are both called "fluids" for obvious reasons. Their commonality stops at mobility and deformation only. In liquid, atoms retain proximity to one another held by the pliant electron web even though their inter-nuclear coupling has been severed. In melting, the electron web must reconfigure itself to accommodate the extra energy that has separated the atoms. As such, liquids are incompressible unless this extra energy is removed first by cooling.

Gas

A gas consists of separate independent atoms with nothing between them but empty space. If an external force is applied, gas molecules are compressible together with heat release and a temperature rise. This is the principle of refrigeration when a liquid like Freon or ammonia is allowed to expand (adiabatically or same energy content) to chill by absorbing heat from a cooling chamber. If the evaporated gas is compressed, heat is expelled and the gas returns to liquid.

Plasma

Plasma is a body of *mobile* positive (+ ve) charge *ions* from gas atoms that have lost electrons that are stripped either by radiations, or electric fields or collisions with electrons. Normally repelling one another by the same + ve electric charge, plasma is "bottled" in a magnetic trap. A plasma often results from a gas discharge. Spectacular Northern Lights (Aurora Borealis) over the Arctic sky is plasma caused by high energy electrons (beta rays) with velocity (V) from the Sun trapped by Earth's magnetic field (B) in transit and spiraling under a (B X V) deflecting torque along its (B) flux lines towards the North pole ionizing air molecules in their paths into plasma. Excited air ions recover by emitting bright colours that make Northern Lights. As gas discharge normally occurs under low pressure (below 100 Torr , or mm Hg), plasmas are uncommon in our daily events at ground level. Plasmas are big players in upper atmosphere where air is thin, *dry* and their *high* energy is responsible from winter climates to Earth's magnetic field as will be shown later in Chapters 7 and 9.

Plasmas resemble metals (solid) but are fluids (either gas or liquid*). While metals use – ve electrons as free charge carriers, plasmas use heavier + ve ions to do the same. Heavy + ve ions provide plasmas their *very* high energy. Both metals and plasmas are good energy absorbers and emitters. Both make excellent Faraday shields as shown by communication blackouts in early space vehicle re-entry. As each proton is 1840 times the mass of electron, plasmas high energy can reach 10 – 50 KeV range, comparable to X-rays. A plasma that glows is called a corona. Plasmas are not limited to just gases. Molten mineral (lava) is a high density hot liquid plasmas with electrons *thermally* stripped. Either as a heavy liquid or light gas plasmas, high temperature in the 2000 – 5000 C range is typical of plasmas. Full of mobile ions, a liquid plasma reacts to an approaching magnetic field with an induced magnetic field to oppose it. This is crucial to understanding how volcanic eruption and earthquake arise.

*Note. Induction heating by RF currents at 450 KHz commonly used in smelting is a good display of liquid plasma properties of molten minerals (magma) by its response to magnetic fluxes. Hot magma flow inside Earth core produces Earth's magnetic field.

25 - 60 Km above ground is a band called "ozone" layer in Earth's atmosphere. Oxygen atoms are ionized into $O3+$ radicals by solar radiations to optically filter out UV-B rays that cause skin cancer. Ozone is plasma and its *daily* absorption of UV-B rays in sunlight heats it up from - 60 C to 0 C. Air moisture has been frozen out in the Troposphere below. One way for ozone band to lose its heat is by collisions with molecules above and below after dusk when solar absorption stops. It is NOT the only way for ozone to cool. There is another more dynamic and catastrophic method to cool when it gets too hot. Details are in Chapter 9.

The Ionosphere between 80 – 200 Km above ground is also plasma. It plays a major role in upper atmosphere climate. Plasma absorption of sunlight here, especially in winter months when solar angle is low and close to the horizon all day in polar regions, can reach 2000 C temperature (though still cold to the "touch" because of its low density). This is a source of high energy. More discussion on this in Chapter 10 about Jet Stream.

Non-matter

Non-matter have no physicality like mass, size, shape, colour, or odor that our senses can discern. They are however as real as matter and detectable by instruments. Non-matter is a force field. Examples are gravity, electric field, magnetic field, light, heat, and all kinds of radiations. Without a mass to accelerate, force fields propagate instantly at the speed of light (c = 3 X 10*8 meters / second or 7.5 times around Earth Equator). A pulsating and *moving* electro-magnetic (E-M) force field delivers radiation energy. Radiation is invisible to eyes like matter is, but its movement is detectable by instruments. I separate "radiation physics" in this mode of invisible field energy transmission from "particle physics" that shows visible matter in motion. Radiation implies <u>frequency, amplitude, resonant coupling and impedance</u> matching. All four parameters play roles in radiation energy transmission, and its reflection.

A hydrogen atom (matter) has a nucleus (proton) with an electron orbiting far away. The empty space between is roughly 220,000 times the nucleus size. This space is void of material, but full of energy field binding electron to the proton. Is it empty ? There is so much to learn about non-matter from this viewpoint. An energy field like gravity, electric and magnetism can be either static and passive (steady with time) or active and pulsating. Energy fields pulsating with time are (E-M) radiations. Field motion is the energy transmission mode under resonant coupling between sender and receiver. That includes the sender, the receiver and everything in between all harmonizing in phase from end to end in energy flow. Slight variation on any party alters the coupling. It is crucial to know that radiation is an energy *extraction* from a field. Radiation energy obtained is demand controlled. In particle physics, each particle once left its source is limited to its internal energy that stays intact in motion. In radiation, the only constant is their common frequency. Others are subject to influences and conditions that may change in transit.

Energy field motion obeys different rules from particle motion. Frequency dictates choices. It picks the medium (called dielectric) of minimum impedance through which the energy field moves. Radiation energy can be steered by manipulating the medium of transmission. Indeed this is why metal wires are used to conduct low frequency current, waveguides or strip lines for microwaves and optical fibres for light. Vacuum of space is no barrier to very high frequency radiation transmission. The higher a frequency, the more sensitive it is to the medium dielectric between sender and receiver. Distance attenuates energy down as the medium absorbs energy along the way. Resonant coupling is paramount in energy flow. Good impedance match facilitates smooth energy flow. Otherwise, it triggers reflection. If it occurs, outward and reflected waves collide to develop standing waves inside the conduit that freezes energy in stagnation. At very high frequencies, component geometry introduces inductive and capacitive effects that are called parasitic.

Radiation waves are invisible to naked eyes and some visualization to their propagation is helpful. Tie one end of a 12 to 15 feet long rope to a door knob, or any fixed point at elbow height. Step back with the open end to a spot where the rope dangles loosely just off the floor. Wiggle rope open end up and down in repetition slowly at first then faster and faster. Notice the weak wave formation at the start but increasingly stronger and flowing away from your hand with rising frequency. At low frequency the medium (rope) through which the waves propagate kills the radiation due to its low energy content. Higher frequency inserts more energy into each wave to overcome the damping to proceed further. Higher frequency raises both acceleration and deceleration rates of its wave angular velocity that are the forces behind radiation. There can be no energy radiation without acceleration and deceleration.

Chapter 3

Atomic Theory, Orbitals, Vibrator, Resonance and Coupling factor

Atomic Theory

The atom has been intensively studied since John Dalton (1766 – 1844). There is an ocean of excellent literature on the topic at all levels, and it is not my interest to duplicate any of it to waste ink on paper.

Danish physicist Niel Bohr (1913) postulated the H atom as a miniature analogue of solar system with the nucleus (proton) at the centre (Sun) and one electron whirling around some distance away (planet). Proton has a mass of 1.66 x 10*-27 Kg., carries a positive + charge of 1.6 x 10* -19 Coulomb, and a diameter in the order of 10* - 5 nanometer. The electron is 1840 times lighter, much smaller, with an equal but negative – charge orbiting about 0.1 nanometer (1 Angstrom) away. Exact numbers are trivial, but their magnitudes relative to each other provides scale for our discussion. It has been pictorially described that nucleus in an atom is in size like a fly inside a cathedral. That is to say that a simple atom like H has 200,000 times more void than matter. It gets worse for bigger molecules. My point is to prepare readers for the vast emptiness in material despite what our eyes tell us otherwise. Human poor visual resolution limit is too low for reliable judgment in small things. Our inability to see skin pores has raised questions about nano-particles deposited on garments and clothing to wear on the body. Inadvertent infusion by them through skin pores (around 50 microns) that are tens of thousands times larger is a real possibility.

* Notes. 1. * denotes power index throughout the text.

 2. A nanometer is $10^{*}-9$ of a meter, a thousandth of a micron long.

Empty space between nucleus and orbiting electrons is vast but void of matter only. There exists a strong electrostatic Coulomb force field within it to pull – ve electrons towards the + ve nucleus. This invisible force field in the atom dictates its behavior far more than the solid particles that define the atom physically can.

Rutherford's Paranoia.

For centuries in ancient Asia and Middle-East, gold has been sought after for jewelry and lady ornaments. Not only does it resist atmospheric corrosions and stays glittery (unlike silver that tarnishes), its extreme ductility and malleability are dreams for craftsmen and jewelers. Pure (24K) gold is soft enough to be rolled or pressed into foils thinner than paper for gilding statutes of worship like Buddha and temples.

In his famous scattering experiment to study the structure of an atom Prof. Rutherford (1908) at McGill University beamed high velocity He ++ (helium) ions, aka alpha (ɑ) particles at thin gold foil targets to map and measure deflection angles from repulsion by the + ve nuclei. His calculations confirmed the smallness of the nucleus relative to the gold atom. Allegedly the emptiness frightened him so much that the following morning upon waking up, Rutherford hesitated to step out of bed for fear of falling through. He earned the 1908 Nobel Prize in chemistry for his discovery.

Coulomb attraction between the + ve nucleus and the – ve electron is inversely proportional to the distance in between **squared.** Pulling force (F) at radius R is down to ¼ (=½.½) F at distance 2R, and 1/9 (=1/3.1/3) F if the distance rises to 3R. Against this attraction, an orbiting electron needs a centrifugal force (=mV*2/R) from its motion in order to remain in orbit. That gives rise to several shells of electrons orbiting at different *frequencies* (ω/2π), descending from the fastest at small radii, R (K, L shells) to the slowest farther out (P, Q shells). Mathematically, a force equilibrium to maintain electron in orbit requires that;

+/- electro attraction / R*2 = mV*2/R Centrifugal force   (1)
On simplification becomes   V ~ √1/R   (2)

It means electron orbiting velocity (V) **slows** down inversely to square root of radius from the nucleus. So does its kinetic energy (= ½mV*2) but at a linear rate with radius. An electron at 2R radius has V√2 = 0.707V of the velocity at 1R away, and half its energy.

That means that outer (P, Q shells) electrons possess less kinetic energy (= ½mv*2)

An object revolving around another in a circle has angular velocity (ω) radians /second. The revolution has a repetition rate, frequency F, of (ω/2π) because each circle has 2π radians. The relationship between angular velocity (ω) to orbiting velocity is via radius (R),

$$V = \omega R \qquad\qquad (3)$$

Rearranging equation (1) gives;
    +/- electrostatic attraction /R*2 = m(ωR)*2/R   Centrifugal force
On simplification becomes   ω*2 ~ 1/ R*3   (4)

It means orbiting frequency F (=ω/2π) increases towards the nucleus at shorter radii, but at a rate inversely to the square root of the radius cubed.

If orbiting *frequency* drops with a rising radius, then outer shell electrons would resonate with lower frequencies in the radiation spectrum such as visible light, and would be excited by 1 – 10 eV (electron volts) energy. This is the case in the photoelectric effect.

Inner shells (K, L) electrons are bound tighter to the nucleus, known as deeper in the energy "well", with higher kinetic energy at higher frequencies. They resonate to higher frequency radiations with higher excitation energy (10 - 50 Kev) X-rays in order to be kicked free. Orbiting frequencies around 10 *18 and above are orders of magnitude too high to be reproduced possibly by any laboratory. 10 *12 is believed highest on record. If these orbiting electrons were observed under a magic microscope, they would seem to be just a haze by moving faster than light photons used to pinpoint their locations, assuming that our visual cortex is up to the task of acquiring and processing the insanely rapid arrival rate of data in the observation. Even the fastest camera made would find it an impossible challenge.

Notes. 1. Frequency is defined as number of orbiting cycles per second (cps) around the nucleus.

2. ω is angular velocity of orbiting electron, which is velocity V divided by R radius, as (ω =V/R)

3. eV, electron volt is energy for an electron to rise 1 Volt, = 1.6 x 10* -19 joule

4. K, L, M, N, O, P, Q refer to shells of orbiting electrons with increasing radii.

Orbiting electron frequency selects resonant coupling from radiation energy. The most dramatic demonstration of resonant coupling was allegedly done by an opera soprano shattering a crystal gobbler with her high C note vocal energy across the concert hall.

Orbitals

Race an elephant with a squirrel side by side on the same round track. The elephant may be bigger and stronger than a squirrel but its size and mass handicap its speed and acceleration as it lumbers on the track. As a result, the squirrel will out-lap the big beast with higher repetitions (frequency). Size and weight pose upper limits on *rotational* frequency around a force of attraction. The race track is the rotation *orbital* for both animals. This applies to molecule as well. Electrons are the smallest and lightest in atoms. They whiz around very fast in orbitals around the nucleus. Their combined pull wobbles the nucleus into its own orbital motion with a *characteristic* frequency set by its mass and perhaps size. At a constant temperature this characteristic frequency stays with the nucleus and would not alter. Like the squirrel and elephant, the nucleus orbital frequency, being thousands times heavier than electrons, will be slower than electrons by same ratio. In solid, the nuclear orbital would likely be *round* if the material exhibits no anisotropic (direction dependent) properties. It is so in order for strength, hardness and other properties to be equal in all directions.

A neutron and its weight moving in a round orbital represents kinetic energy of motion only. A proton with a + ve charge on the same orbital is a different matter. When any + or − ve charge is in motion, a concentric magnetic flux encircles its trajectory. If motion is a closed loop, the magnetic flux spirals around the loop like a toroid to create a pulsating N-S magnetic dipole perpendicular out of the track plane, at a *frequency* of the rotation. The N-S magnetic dipoles of orbiting protons couple each other *in close proximity* (orders of Å, Angstroms) as magnetic bonds. Try to pry apart two magnets stuck together to feel this strong force.

If a solid is heated, resonant energy is coupled to the slowly orbiting nuclei. Electrons are too fast for heat coupling. Extra kinetic energy elongates the nuclear orbital to accept the extra momentum. This brings volumetric expansion. With more heat added, magnetic bonds between nuclei are strained and can rupture. Bond rupture means Melting Point is reached. Melting turns a rigid solid into a pliant flowing liquid held together weakly by the pliant electron web.

Vibrator

After melting and with more heat, nuclei orbitals continue to stretch longer elliptically as fluid with tiny thermal coefficient of expansion around 0.1% per degree C. If heating continues the nuclear orbitals back and both become so linear that it is like shaking. They are now vibrators building up kinetic energy to break free as gas molecules, a process called vaporization. Vibrating gas molecules generate vapour pressure by repeated collisions with the container wall.

Resonance

Resonance is in our daily life more than we realize, in events of a cyclic nature with any rhythm. A parent visiting a local park with children for the swings applies resonance automatically. With child in the swing seat, we push it from behind or in front repeatedly in synchronization with the swing to keep it swinging. That *IS* resonant coupling between energy sender (parent) and energy receiver (seat and child). Frequency (number of times per second) and timing (when to push) have to synchronize between the parties. This requirement *must* be met for radiation energy to flow smoothly. Resonant coupling is delivering *pulsating* energy in lock-step and in phase together.

Tuning to a radio or TV station is also resonant coupling. Pulsating (E-M) signals from the stations already exist in space but without delivering energy anywhere *until* a receiver is tuned to it. Resonant coupling is energy extraction from an energy field by the receiver like opening a water faucet. There is constant water pressure but no flow until the faucet is open. In radiation energy flow, *demand is always the control* and not the supply.

Impedance

In DC (direct current) and low frequency AC (alternating circuit) circuits, energy flow is in the current (I). Current (I) is driven by a voltage (V) from a power supply to overcome the circuit resistance, R.

At high frequency, complications arise from two other frequency sensitive components, called reactance R. One is a <u>magnetically</u> sensitive inductance, L $(R=\omega L)$ that generates a voltage to oppose current (I). L is due to leads, wires and coils. It rises with frequency (ω) obviously to suppress current (I). Second is the <u>electrically</u> sensitive capacitance, C $(R=1/\omega C))$, that behaves opposite to L by dropping R value with frequency to enhance current (I). In a high frequency RF circuit, the combined reactance (ωL), $(1/\omega C)$, and R (resistance) total is called the "impedance" (load) of the RF circuit. It determines the current (I) by how well it matches other impedances R in the RF circuit.

Apart from resonant coupling, output impedance of power source and input impedance of receiver has to match well where they meet each other for smooth energy transmission, like two pipes at a joint for liquid flow. If not, RF power is reflected by the receiver that sets up standing wave in a collision between outgoing and reflected energy to result in a stagnation.

Coupling factor

Resonant coupling requires frequency match within a narrow range between sender and receiver. Searching for a radio station, signal is strongest when station frequency and radio receiver tuner agree exactly. Any deviation weakens the signal. The degree of coupling limits energy transfer level (amplitude). An example is solar absorption by various shades of darkness on the absorber. The darkest has the highest coupling factor. Interference lowers coupling factor, eg. operating a receiver in bushes or behind a building out of line of sight.

In circuitry terminology, coupling factor depends on "impedance matching" between sender and receiver. The closest physical analogue to impedance matching would be pipe size diameter in a liquid flow between adjoining sections. If the two pipes are same size, then liquid flows smoothly and uninterrupted. If sender pipe has larger diameter than the receiver pipe, liquid will gush out in an overflow at the joint. Conversely, if sender pipe is smaller than receiver, liquid flow slows down after the joint. In radiation resonant coupling, a sender lower output impedance meeting a receiver higher input impedance triggers a big reflection. The opposite triggers an increased energy flow on coupling. Optically it is how we see colour of an object in either reflection or transparency by the degree of coupling and matching.

It is inferred that for efficient radiation energy to flow, two independent factors must work in tandem and in harmony;

1. a tight frequency agreement for resonant coupling, and
2. a close impedance match to minimize reflection at coupling.

Chapter 4

Energy classification, E-M mechanics, Parasitic and Radiation Law

Energy classification

What is energy? The first fire, perhaps by accidents of a lightning, introduced humans to light and warmth (heat). Along came Stone Age and old timers learnt to make sharp tools by striking one shale against another in order to bludgeon or impale enemy in fights (kinetic energy). Bronze Age taught them how to use *blast furnaces* to mold utensils and make more tools. Fast forward to James Watt's (1765) steam engine that used heat for water vaporization to propel train locomotives. That single invention was responsible for Britain successfully building the vast British Empire in India and Africa for two centuries. It enabled her on one hand fast military deployment to defend her colonial interests and on the other rapid repatriation of precious raw material from colonies to home market for obscene profits until it was replaced by Rudolf Diesel and his superior invention. The success was later repeated in America linking two coasts from East to West in like fashion. Faraday, Edison and Tesler ushered in electricity to light up a bulb and to turn motors for fan and pump water. World War 2 ultimately debuted nuclear power that stunned the world speechless by unleashing hell in Hiroshima and Nagasaki of Japan that won an unconditional surrender. Application of energy has been a steady companion to civilization all along for eons. Energy applications surged quickly under profit and strategic inducement. Understanding energy as a science has however not kept pace with each new discovery. It is a classic example of **technology** leaping ahead of **science.** There is always grave danger in being unaware of potential negative fallouts and nasty surprises like marketing a new life saving medicine without adequate study of its potential harmful side effects.

Dictionary defines energy as "ability to perform work*". That hardly explains its makeup nor nature. Scientifically, it helps to know that a water molecule has two hydrogen atoms linked to an oxygen atom in between. Water then has a structure and a symbol as identity. Energy still lacks that even as late as we are now in the 21st century. This book is an attempt to fill that gap for you. Armed with such knowledge, we should be better able to predict accurately consequence of energy use, develop an ethical attitude and moral mindset to it as part of our tools to make responsible decisions for our actions, and as guidelines for future generations.

* Work = Force X Distance   (moved in the direction of the force.)

Energy requires two components for it to exist, the first is a **force** (can be steady or time variant) and the second **its motion.** Either one without the other cannot develop energy. According to Newton's (Second) Law of motion, there would be no motion without a force. Go one step further, there is no energy **unless its force is in motion.** Before we go any further, some old misconception needs correction as well. It has been misquoted for a long time that "energy can neither be created not destroyed". It is untrue. Energy can only be released from storage, and therefore is not created synthetically. Energy can be, however, **neutralized or annihilated.** Destruction is a matter of semantics. All it needs is cessation of motion (based on above statement) for energy to die. When two ball bearings of same material and equal speed but in opposite directions collide, an **elastic** encounter (defined by **no** energy loss) would send them back in reverse directions after impact. In reality all collisions are **in-elastic** with some energy loss. In this more realistic scenario, both ball bearings would stop dead after impact. This is an example of two equal but opposite kinetic energies cancelling out each other with their energy dissipated at once.

Energy can exist in both **matter** and **non-matter** domains. In the matter word, the force would be an actual physical pressure to push an object in motion that is either linear (straight) or curved. For rotary motions, a torque replaces force, and rotation is the motion. Both are **kinetic energy.** Potential energy is negative – ve kinetic energy in reverse, or in stationary

storage. It is energy supplied by a superior force pushing against the original force (eg. gravity) and the distance covered is in a direction opposite to where original force intends to go. Examples are moisture raised up by sunlight to become clouds, a steel spring compressed, a sling shot or elastic band or bow stretched before release. Energy storage comes from inter-molecular magnetic coupling reconfiguring itself to accommodate the energy injected. Potential energy is equivalent to heat. If a gas is compressed (less volume), heat is expelled under thermodynamic laws. If this heat is captured and stored, it is same as potential energy of an object held high in a gravity field.

In a matter world dealing with objects, energy is a physical force in motion. In a non-matter word, energy can only exist as a pulsating (E-M) radiation force field in resonant coupling between sender and receiver in a transmission. We can segregate energy into its various forms by frequency, from the fastest to slowest.

1. Lethal Radiation.

Extremely high frequency sinusoidal (E-M) force fields pulsating transversely to its direction of propagation in resonant coupling between a sender and a receiver at the speed of light, c. This spans a spectrum from ($10*18$ to $10*25$ Hz) (100 KeV–100 MeV) gamma (γ) rays from inside a nucleus, ($10*17$ to $10*21$ Hz) (1 – 100 KeV) X-rays from electrons in deep K shell of a molecule, to ($10*15$ to $10*17$ Hz) (10 eV – 1 KeV) invisible Ultra-violet (UV) light that is enough to ionize a molecule.

2. Visible Light

This is just a thin band of frequencies ($10*14$ to $10*15$ Hz) (1 – 10 eV) from violet to red in the visible spectrum. Energy is enough to trigger electron transition between shells.

3. Heat

This spans from ($10*12$ to $10*14$ Hz) (< 1 eV) Infra-red (IR) just below visible light through EHF millimeter wave band that is reserved exclusively for radio astronomy and down to microwaves to transmit long distance telecom signals. Energy resonates closely with nuclear motion and slow electron transitions.

4. RF radio

This covers both FM and AM radio signals, and applications of induction heating in smelting

5. Sound (audio)

In air and water as energy transfer media, pulsations are not transverse but longitudinal compression waves in the same direction of propagation. This is the bottom end of the frequency spectrum in air. The energy is acoustic including ultra-sound imaging used in medicine and seismographs in resource exploration. Each energy quantum is called "phonon".

6. Sub-audio

Under-water transmission mainly used by navy for submarine telecom and sonar echoing by fishing industry.

7. Electricity

This energy can flow as either AC (alternating current) or DC (direct current). Energy flow is a pulsating (E-M) *force field in motion* within a conductor, but NOT of electrons. This argument is based on its instant acceleration to the speed of light, $c = 3 \times 10^{*}8$ meters /second or 7.5 times around the Equator. Only a force field without mass can achieve it.

8. Kinetic

This is impact energy where a physical force or pressure (expanding gas in an explosion) delivers motion. Motion can be one direction (straight or wavy in varying directions) or rotary (in flywheel). As is obvious, kinetic energy sits in the bottom rung of the energy hierarchy. If an object moving under kinetic energy is stopped by impact that is the end of energy. No motion, no more energy. From kinetic energy to create other energy forms is an uphill struggle. Reverse is easier. Light becomes heat with over 90% efficiency using black colours. Likewise, electricity runs motors to create rotational kinetic energy efficiently

9. Potential

Potential energy is kinetic energy in reverse, or storage under tension provided by a superior force. It is represented by the return half cycle in a simple harmonic motion (SHM)

10. Nuclear

I consider all fission materials a private reserve of Nature and at her exclusive disposal only. It is a non-renewable family heirloom for all generations across time. I oppose its unethical exploitation for a selfish short-term benefit but leaving long-term nightmares to our children. Until a reliable permanently safe waste disposal solution is found, leaders of conscience must stand pact against its proliferation except in medical treatment. To be consistent with our claims of loving and protecting our offspring, it is sheer hypocrisy pronouncing one thing but doing exactly the opposite

E-M Mechanics

Of our five sensory organs, the tongue and nose work chemically. The eye and ear both work with frequencies but at different ends of the spectrum. Ear responds only to audio-frequency (AF) from 20 Hz to 20 KHz by *air* vibrations. In vacuum or water, we cannot hear sound. Eyes respond to extremely high frequencies between $10*14$ and $10*15$ Hz.

Energy flow works differently at different frequencies. Lower frequencies need matter in the energy transfer and are therefore *matter* waves. In matter waves, material movement is the energy flow. Time necessary to accelerate mass limits frequencies up to a ceiling of $10*11$ Hz. Higher frequencies employ pulsating (E-M) force fields without mass motion in its energy transfer. They are *radiation* waves that can propagate through any medium including vacuum.

Wave mechanics is a study of (sinusoidal) wave motion and energy flow whether it is a tiny pebble dropped into a pond or a huge tsunamis caused by an underground tectonic shift. Visible waves on the surface can be mathematically modeled for prediction of their propagation. Where a wave is moving, there is an energy flow. On contact, *matter* waves are reflected with both material and energy flow redirected towards a new direction.

With *non-matter radiation* waves, the pulsating E-M force field motion carries the energy flow *under* resonance coupling with the receiver. There is NO material motion in its energy propagation. A force field may spread out in all directions but energy flow does not copy the same pattern. Only resonant coupling by a receiver can initiate force field motion and energy flow. If reflection occurs with E-M force field, there is no redirect like matter waves. When a pulsating E-M force field collides with its own reflection from an impedance mismatch, a ***standing wave*** results and energy transmission will stop in stagnation.

At any given time, there is an ocean of radiation waves with pulsating (E-M) force fields crisscrossing one another in space, but there is no energy flow to anybody except to the only receiver that is in resonant coupling to a particular signal. In the water analogy, there is water pressure throughout the system, but water flows only to the open faucet. Likewise, the burning hot sun is not radiating out sunlight in particles (photons) to far corners of the universe wasting thermonuclear fuel. Solar energy only goes to stars that intercept its pulsating (E-M) force field in resonant coupling.

Engineers who work with radio frequency (RF) circuits or waveguides must deal with reactive components such as inductor (L) and capacitor (C) in the circuit that limit the current flow. At low frequencies, these reactive components are negligible and resistance, R is the only impedance to energy flow. Non-matter radiation waves are at very high frequencies, and reactive components (L and C combined) dominate over resistance R in the circuit impedance. Sunlight is a conglomerate of many very high frequencies of radiations.

Parasitic

RF circuits suffer from energy leakage to structural imperfections. Collectively called parasitic, its origin is due to how components are made. Incidental inductance (L) and capacitance (C) parasitic that can be ignored at low frequencies start interfering at higher frequencies due to (ωL) and ($1/\omega C$) effects. Once in coupling, parasitic siphons energy from sender. Increasing parasitic with rising frequency is like changing to darker clothing to feel warmer in the sun, giving wearer a false impression of a hotter sun. Solar intensity may stay constant but a higher coupling factor due to better impedance match raises input. It is equivalent to the voltage of a circuit being constant but impedance has dropped to let a higher current flow. It proves that in radiation energy, demand is in control of power flow.

Radiation Law

One good way to understand radiation physics is by a scientific definition. Radiation is a self-contained radio frequency (RF) circuit that links a power supply (transmitter voltage) at one end to a load (receiver) at the far end by a conduit (or path) through which a RF current (I) flows. There is NO separate return path. In a waveguide or an optical fibre, the same conduit conducts RF energy both ways. In air or space, the conduit is either a direct beam (line of sight) or reflection bouncing off the ionosphere in CB (citizen band) and HAM amateur radio chitchats. As in any RF circuit, circuit impedance governs RF current level. Solar radiation meets the definition of a RF circuit with a proviso. Its wide light spectrum makes it an ultra-broadband multi-frequency radiation and not a monochromatic (single frequency) RF circuit. Sunbeam is like an ultra-broadband optical fibre, but in space.

Radiation energy is ***non-matter.*** Radiation energy is pulsating (E-M) ***force field*** in ***motion***. It happens as a last resort where there are no other viable alternatives to deliver a high energy flow. It carries a high price tag. Hardly effective below 200 C (the price tag), but powerful once initiated. Its rate of output obeys a fourth power law of absolute temperature called Stefan-Boltzmann Law of radiation ($J = \epsilon.\sigma.T^*4$, where ϵ is emissivity of radiator (1 for ideal blackbody), σ is Stefan-Boltzmann constant, T is absolute temperature in degrees Kelvin (= radiator temperature in C + 273) and J is joules per second (or watts) per square meter of area of radiation.

The best way to illustrate Stefan-Boltzmann Law governing ***blackbody*** radiation energy is by a real example. Human body is at a stable temperature of 36.7 C. Skin temperature is more like 33 C to the touch. Clothing screens it down to the 25 – 28 C. With over 80% high water content, our body is a radiator. Water emissivity, a measure of heat emission ability relative to blackbody of 1, is 0.998. At 25 C, a 6-ft tall body weighing 200-lbs radiates (***arguably, assuming no conduction or convection that are far more efficient***) away 80 to 100W, or at a rate of 40 to 50W per square meter. This energy loss needs to be replaced constantly by food intake and activity to prevent hypothermia that can cause death.

Planet Earth stays at a comfortable 15 C, less than half of human body temperature of 36.7 C. Over 72% of Earth surface is water. Majority of remaining land is covered by vegetation that also has high water content. So Earth emissivity is close to a blackbody. The following table compares three neighbouring planets in the solar system in how much solar heat they "radiate" away without adjusting for coupling factor differences from different atmospheres. Only Earth has water.

Properties	Venus	Earth	Mars	
Distance from Sun (AU)	0.7	1	1.5	(AU is 93 million miles between Earth and Sun)
Solar irradiance, watts/m*2.	**(2000)**	**1000**	**(444)**	(computed unadjusted)
Temperature in Celsius	460	15	- 50	
Temperature in Kelvin	733	287	223	
T*4th power, 10*8	2887	67.85	24.73	
Stefan-Boltzmann const, σ	5.67	"	"	
Radiation output, watts/m*2	16396.3	*384.7*	*140*	***computed outputs***

This table illustrates that Venus at 460 C is able to radiate out 100% solar heat received, but Earth and Mars at 15 C and − 50 C respectively are delinquent by a factor of about 3 in their ability to radiate away solar heat received. For every 100W of solar heat absorbed by Earth, 62 watts require a more efficient method of disposal to maintain temperature equilibrium. We will see how it is achieved later in Chapter 9, Heating, Cooling and Avalanche Gravity Bernoulli Cooling.

It must be made clear that Radiation is NOT particle physics. There is confusion among even scientists. Photon is just a mental construct used to explain Einstein's photoelectric effect. Photons do not exist. They cannot have mass in order to reach instantly the speed of light (c = 3 X 10*8 meters/ second or 7.5 times around the Equator/ sec.). No mass is no physical existence of a particle. Radiation is a pulsating (E-M) energy field in motion. Gravity is omnipresent but a stationary passive non-pulsating field. We are literally swimming in a myriad of pulsating radiation fields and a gravity field at all times without experiencing them consciously. No radio will play nor a cell phone will ring until resonant coupling happens. In seeing a bright red rose, red colour wavelength in sunbeam has coupled resonantly with the rose but the excess is reflected after saturation at that wavelength. No other colours appear because they are either all absorbed below saturation or fail to couple resonantly.

Chapter 5

Heat. Water, Tug of War between
Sun and Earth Gravity

Heat

Strike a rock hard with a hammer, the rock cracks because it is not ductile and kinetic energy from the blow has found and ruptured weak molecular bonds inside. Rocks are insulators that present high resistance to energy flow. Do the same to a ductile metal, fast deformation takes place to avoid rupture and molecular bonds have been reconfigured.

Metals conduct both heat and electricity. It means that an energy *field* can penetrate its bulk with little opposition made possible by a pliant pervasive electron web that weaves through between atoms like a honeycomb and enclose them inside a skin. Energy absorbed is **evenly** distributed throughout its lattice instantly at the speed of light, c. High metallic ductility and conductivity come from this uninterrupted pliant electron web that distributes energy with equal shared disruption throughout to all atoms.

Heat is activity at the invisible atomic level. Heat is generated by atoms **reacting** to a pulsating (E-M) force *field* passing through. High heat emission carries its characteristic temperature profile plus high and low limits like a wave with a peak frequency that can be read optically by a pyrometer. Low temperature heat at 100 C from boiling water may peak at ultrasonic frequency around 250 KHz. Temperature peak of 400 C, enough to melt tin or lead for circuit soldering, is up to 350 Giga-Hz that is just below infra-red (IR). In a low red glow, peak frequency is nudging the low end of visible light close to 10*13 Hz. Temperature profile peak *frequency* shifts exponentially fast with rising temperatures.

Heat absorbed excites atoms into higher **frequency and amplitude** by shaking. Increased activity stretches inter-atomic separation to bring volumetric expansion. As more heat is injected inter-nuclear (or atomic) bonds continue to stretch by shaking until they rupture and sever finally to begin melting. This is the point of phase change from solid to liquid. In a liquid, the pliant electron web provides the only weak cohesion to retain the atoms together. Its pliancy allows the liquid to deform, to flow and to be separated easily.

If inter-atomic bonds are sufficiently strong to defy stretching for volumetric expansion, injected heat poses such a strain on the bonds that the agitated atoms raise their shaking frequency (when more amplitude is prohibited) faster to seek relief desperately. Temperature shoots up rapidly in a struggle to radiate away the heat. Higher frequency raises acceleration and deceleration rates of nucleus vibration to enhance radiation. Recall the wiggling rope analogy. This is a clear inference that inter-atomic bond is magnetic in nature that can be weaken by heat until it reaches Curie temperature when all magnetism is destroyed.

Water (H2O).
This quintessential liquid for life in both fauna and flora is the pinnacle of nature's marvel. Without water, life is impossible under most circumstances. Humans may last for weeks without food intake, but only days without water. Dehydration is a ruthless merciless killer. Water conducts heat to organs in our body for temperature, and conducts electricity to flow in our nerves as signals.

Water is among the densest and heaviest liquids (*Figure 2*). Its high density (1000 Kg / cubic meter maximum @ 4 C) delivers impact power in waves that crash ashore and erode shorelines. Storm surges in front of a hurricane or tsunami is responsible for major property damage and its retreat is even deadlier by sucking objects, people and animals who may have survived initial frontal onslaught back out to sea to drown. As a vapor however, water is the fifth lightest only after hydrogen, helium, methane and ammonium. Warming up from 4 C to boiling at 100 C, liquid water expands at a stunning ***4.2 %*** thermal expansion coefficient over 96 C when < 0.1 % is typical for most liquids over the same temperature range. It is **4200 %** more than other any liquid. It suggests that water molecules are low tensile, pliant to heat and elongates easily at will. Such exceptional elasticity and high temperature expansion coefficient of water molecule accounts for its effectiveness in convectional cooling. Density change is tied closely to thermal expansion coefficient. High density change means strong convection power in Archimedes Principle. Water volumetric expansion from liquid into vapor is equally phenomenal at the incredible **1500 – 1700*** times depending on atmospheric pressure. This property formed the backbone of steam engines that ruled global commerce from 18th to 20th century until replaced by Rudolf Diesel (1893) and his superior engine. Moisture in air continuously absorbs sunlight to expand and rise higher until it cools to condense as reflective white cloud (hexagonal microcrystalline snow) that stops further absorption.

*Note. Dry air density is 1.1 Kg/ cubic meter. Its molecular weight averaged over the (78% N2 + 22% O2) mix is 28.8. Water (H2O) molecular weight is 18(=16+2). Moisture is 62.5 % as light as dry air. By ratio, moisture density is 0.6875 Kg/cubic meter (= 1.1 X 18/ 28.8). Expansion of water from liquid to vapour is therefore **1454.5** times (=1000/0.6875).

Earth gravity pulls all things down as hard as possible. Moisture being lighter than air is hence;

1. **a rebel to gravity always rising above air in defiance to escape, and an even**

2. **bigger rebel when moisture absorbs heat to expand and gets lighter still.**

Knowing these two facts will help you understand how weather works in later discussions.

Water has a unique anomaly as well. At **4 C** just before freezing water reaches m*aximum* density. It reverses direction of thermal coefficient of expansion at 4 C as shown in **Figure 2**. Water **expands** from 4 C to 0 C before freezing into ice. As a result, ice is 92% of water density making it float on top as icebergs do, hiding 9 times its visible size below as a menace to maritime traffic. It is also the principle of skating where the high pressure of steel blades compresses ice to melt into a thin water layer for the skate to glide on. This property is very important to marine life in winter. Ice cover over lakes and ponds is a cold shield to prevent further freezing below 4 C under ice. On chilling to 4 C at its highest density water below the ice interface sinks towards bottom to start a down convection. At both N and S poles, this action under the ice caps is responsible for strong submarine ocean currents to stir up bottom nutrients. Its extraordinary 4.2% over 96 C thermal expansion coefficient drives a strong convection under the oceans.

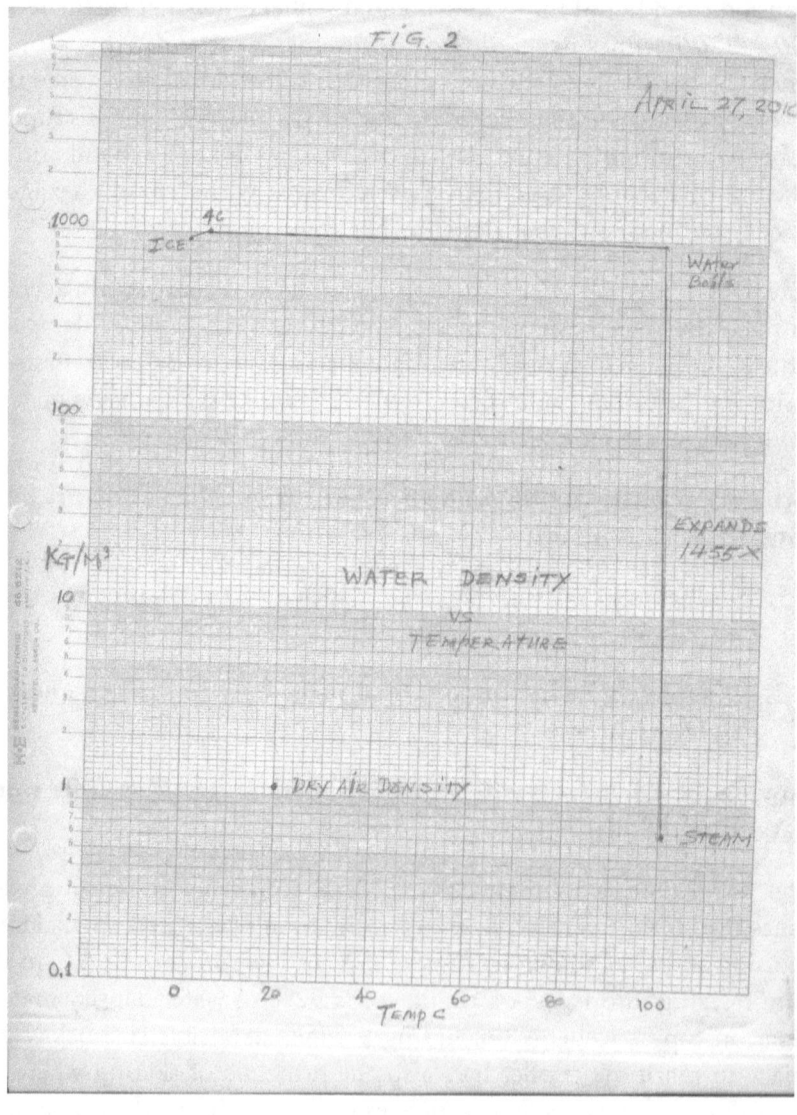

Figure 2. Excel Graph of Water Density vs Temperature

The ability of water to absorb radiation energy is demonstrated by the difficulty in operating any communication device in the bush. High humidity acts as an electrical shield. Stretching water molecules apart may be easy (remember 4.2 % ?) but severing them is much harder. Its high elasticity comes from easy bending and flexing like a molecular gymnast, but severing water molecules apart thermally requires breaking a strong +/- electrostatic attraction from its unique +/- electric dipole that relates to its high surface tension of 73 dynes/ cm. Water becoming vapour (evaporation) must climb a high latent heat of vaporization barrier of 539.1 gm. calories / gm. that explains its high surface tension as above. Snow is in reality hexagonal air bubbles trapped like balloons by an outer coating of frozen water. Air bubble size depends inversely on the rate of freezing, bigger bubble for slower freeze and vice versa. Slow freezing allows more air trapped inside each bubble during wet air struggling up under sunlight against gravity to become *white* clouds that have high internal light reflections. Fast freezing occurs in sudden collisions between hot humid air and cold dry air to produce *dark* clouds in which air bubbles are small with less internal light reflection. Fast winds in tornado and hurricane have dark clouds by rapid condensation. Friction from the collision deposits – ve and + ve electric charges to top and bottom of the (dark) clouds respectively that create lightning flashes and thunder claps. The same fast freeze condenses moisture as hailstones to be flung out centrifugally by the tornado vortex as it hops and dances hither and yonder. Fast freezing converts latent heat of warm humid air instantly into kinetic energy to create wind vortices and hailstorms. That is how slow freezing white clouds hold no +/- ve electric charges for thunderstorms, but fast freezing dark clouds do..

Since air is an insulator, snow being full of air bubbles is good insulator too. This explains how an Eskimo igloo remains warm inside and why northern folks purposely pile snow high outside rural homes to block heat loss in cold winters to save fuel. In snow piles, gravity compresses and squeezes air out of bubbles at the bottom layers to turn snow into ice. Ice loses its reflective whiteness to become clear as water. Glider pilots seek and steer towards dark cloud shadows on ground in order to ride warm air lifts called *thermos.* Dark cloud shadows absorb sunlight faster to heat up air above them more than bright surroundings.

Air is less dense than water. This gives total internal light reflection of air bubbles in fluffy snow its reflective white looks. Long term looking at shiny white snow without eye screen causes blindness like staring at the sun. Eskimos wear home made whale bone goggles with a horizontal slit that polarizes reflected light like sunglasses to cut its intensity. The smaller the bubbles, the less internal light reflection and whiteness drops. Ice is 8 % lighter than water and therefore floats on top hiding 92 % submerged below.

Water has a high specific heat that is twice of most common oils. Specific heat is a measure of heat needed to raise temperature 1 degree C. High specific heat of water is indicative of the elasticity of water molecule to stretch like a steel spring to store energy. Its surface tension of *73 dynes/cm.* is highest among liquids due to a +/- bond between molecules mentioned above. It enables insects like mosquitoes to literally walk on it. High surface tension helps drive water up by capillary action inside xylem fibres of plants to deliver nutrients to leaves in tandem with osmotic pressure at the roots. Solar evaporation of water off leaves activates and sustains the elevation process all day. These *two* push-pull natural forces combined lift skinny columns of water against gravity to the tallest Douglas fir in California Redwood Forest and the spectacular giant old growths of Amazon rain jungle in Brazil. There seems no limit to the ability of water to perform miracles. Some folks wonder if plants can think. Evidently they do, or is the design done for them by a superior being ?

Structurally water is a lightweight (MW 18) simple molecule consisting of one lonely oxygen atom placed midway between two hydrogen atoms in the H--O--H configuration. The O--H arm is just under 1 Angstrom long. Why is water so dense and heavy is a mystery that deserves scrutiny. For reasons yet still unexplained electrons whirl around one particular O--H arm more than evenly over both. This sets up a net − ve charge over the O--H arm offset by a net + ve charge at the lonely H at the opposite end. This is called an +/- electric dipole. Electrostatic pull between these opposite charges bends the molecule into banana shape with a 104 degree angle between the two O--H arms. Electric dipoles can resonate with a band of pulsating electro-magnetic (E-M) fields of radiation. In resonance, the water molecule flexes its O--H arms like a bird in flight. Flexing stretches the molecule longer to explain its high thermal expansion coefficient of 4.2 %. The electric dipole can also tumble end over end or gyrate side to side to different radiation frequencies. This makes water molecule the best molecular gymnast there is. Indeed water absorbs sunlight over most of its entire spectrum far more than any gas. CO_2 solar absorption is puny by comparison. Water is the only ONE true greenhouse gas (GHG), but I detest the moniker. I prefer water be known as ***Elevator*** gas for its ability to raise ***and*** lower air temperature by its specific heat, and its ascent up into sky as clouds then fall back down as rain or snow.

Water is so hungry for heat that nurses dab wet cloths to patients to bring down fever and marathoners douse cups of it over themselves to cool while running. Water hogs heat any chance it gets. Once combined, it takes a Herculean effort to separate them. One way to split them up is by condensing or freezing the water out while the heat is converted ***fast*** to kinetic energy in the process. In a respiration analogy, heat is inhaled but kinetic energy (not heat) is exhaled as its replacement. Examples are hurricanes and tornadoes. Heat is the energy source and moisture is its carrier. At the apex of activities, rain and hail are condensed out of the moisture with strong winds carrying the kinetic energy derived from the heat before. In high speed wind, temperature plunges to further condensation or freezing. So heat and moisture couple together by the sun, but split into water and strong wind under the force of gravity.

Heat and ***moisture*** are Nature's lovers. In liquid state, close proximity allows water +/- ve electric dipoles to electro-statically link up hexagonally in multiples of six molecules (MW=108 or 6 X18) into fishnet lattices that act like a heavy molecule with low agility to move or to spin. The familiar six-point snowflake is well known. Moisture freezes into micro-crystals to form white clouds during its slow ascent up the sky while cooling adiabatically. The hexagonal fishnet lattice explains its high surface tension (73 dynes/cm), high specific heat, and high latent heat of evaporation (539.1 gm. calorie /gm). A heavy liquid water lattice resonates to low (sub-sonic) frequencies. It is why sonar works in water and why the navy and whales use it in submarine communication. Sound travels **4.5 times** faster in water (being **909** times denser, =1000/1.1) than in air as all lakeside cottagers can attest to. As a result, objects with a high water content like most animals are susceptible to harm by persistent exposure to low frequency vibrations coming from slow tumbling windmills and 50 - 60 Hz (E-M) radiations of high energy utility power lines. A minimum separation of several miles is recommended to attenuate the impact.

Tug of War between Sun and Earth Gravity
Chapter 8 is a discussion of Archimedes Principle where he found a way to determine the density of an object relative to water. His discovery

annoys gravity. When air gets hot and expands, it loses density to rise up in the atmosphere to defy Earth gravity. Water with its lighter molecular weight (18) than air (28.8) is already anti-gravity to begin with. Affinity of moisture to heat further aggravates the anti-gravity defiance. At first opportunity, Earth gravity would do its best to strip heat from moisture into condensation to bring moisture down.

Wind is air with kinetic energy. Blowing over open water, that kinetic energy churns up waves and storm surges with devastation power. If temperature is cold enough to freeze water, the storm becomes a blizzard that abrades like a sandblaster. A blizzard blowing against glaciers chips it away *faster* than heat melting. It is a collision of ice on ice. Ocean waves crashing and eroding shore lines is how unabsorbed solar heat energy turned into kinetic energy by gravity in its disposal.

Daily there is a titanic tug of war between a hot sun that evaporates surface water of oceans and lakes into moisture to lift it up skyward against Earth gravity. On its way gravity fights back by a process called adiabatic cooling (at the rate of -3 C/ Km) to chill the rising warm moist air until it freezes into reflective white ice crystals as clouds about 5 Km high. Once frozen Earth gravity has won first round by stopping and blocking more solar absorption. Effectively this UP process is tantamount to inhaling solar heat by moisture. This daily sparring between sun and Earth is vital to life that you shall see. It is the source of our nourishing rivers when clouds turn into rain during monsoons. After clouds have fallen down as snow that becomes ice and glaciers, they act as buffer against global warming but also more importantly as energy source to drive deep ocean currents to keep a stable 15 C ground temperature for Earth as will be explained in Chapter 12 – Global Warming Threat Assessment, etc. Ice and glaciers are Earth's thermostat and its temperature reference.

Sunlight is an enthusiastic match maker of heat to moisture. Earth gravity acts as spoiler to separate them. Sunlight is an ultra-broadband optical beam in space between sun as sender and Earth water as a voracious receiver. Earth water therefore acts like a capacitive load. Its net value sets the coupling factor and the level of energy flow. Just like darker clothing in sunlight makes the wearer feel warmer by raising coupling factor, fossil fuel burning that raises air humidity can create a false impression that the sun has become stronger on its own.

Although the sun overpowers Earth, but gravity comes out ahead at the end of each day when sun gives up its fight. That is why monsoons, tornadoes, blizzards, torrential downpours, and storms rarely start in bright sun but occur often at dusk or later into the night, depending on how much residual solar heat stays in the atmosphere when sun goes down. "Red sun at night, sailors' delight" is a mariner expression that relates to low air moisture that allows long red waves of sunlight to come through bright. Low air moisture means low heat content and little chance for storms. "Red sun in morning, sailors take warning" is the second expression, which means that air is dry and thirsty for moisture. It gets plenty of help by a hot sun during the bright day ahead. Large moisture has high heat content with good storm potential when Earth gravity splits them up from each other.

There is a corollary between breathing and solar heat. In respiration, oxygen is inhaled and sent to organs and muscles to burn with glucose to create heat for body activities. In the end carbon dioxide is exhaled. The daily solar function behaves as follows; Refer to *Figure 5* in Chapter 9

1. **Day, Earth *inhales solar heat energy* to evaporate water up the sky as clouds**
   – the UP process.

2. **Night, Earth *exhales kinetic energy* as rain, snow, wind or waves to expel heat**
   - the DOWN process

Chapter 6

Thermodynamics, Temperature, Humidity, Tornado and Hurricane

Thermodynamics is literally heat action. It is a statistical study of gas activity in response to heat. The effect of heat on solid is limited to melting when inter-atomic bonds rupture, or radiation if the bonds held their ground. Heat on liquid lowers its viscosity. Heat on gas is volumetric expansion with higher kinetic energy in the atoms. The mathematics can be daunting but we are not concerned with it here. Our interest is in how heat becomes kinetic energy to produce actions. Heat is an energy field by nature like radiation and its action on gas molecules is seen by increased activity. Oversimplified, thermodynamics is expressed by an ideal gas equation called Charles Law below;

PV/T = constant, where P is pressure, V is volume and T is temperature in degrees Kelvin (= Celsius plus 273).

It states that if pressure stays constant, temperature would vary opposite to volume. This is the principle of refrigeration. To cool a gas like ammonia used in a refrigerator, let it expand by evaporation. An expanding gas sucks heat from surroundings. The opposite is equally true. Heat up a gas by squeezing it. Performing both functions in series is called a Carnot Cycle that turns heat into mechanical energy for work. A good example would be the steam engine used by a locomotive to pull trains.

Temperature

Temperature is in our lives constantly, from our body heat to coffee, tea or soup, comfort level in our homes or offices, car engine, computer, etc. The list is endless. Temperature measures atomic ***activity level*** or ***kinetic energy per unit volume*** (density). Deserts are hot because so much solar heat is jammed into a tiny amount of residual moisture there is, making each water molecule very agitated. Hotness is from ***scarcity*** of moisture to carry the heat load, not from ***the sand.*** A 100% dry desert would be cold like N and S poles where all air moisture is frozen out by extreme cold. After sundown, condensation begins in deserts that leaves a thin morning dew to sustain life. At 20 C room temperature, an atom has 0.025 eV kinetic energy that corresponds to 50 um (micron) radiation or frequency 6 X 10*12 Hz.

Low temperatures under 150 C can be measured safely by a cheap glass thermometer. If access is impossible like a hot furnace at the far end of a factory, and for temperatures into 2000 C or beyond in smelting, a "pyrometer" is used instead to read temperature optically. Looking through a view port and a lens focused at the hot object to be measured, operator adjusts a knob that changes the current sent through a tungsten wire in a sealed glass ampoule same as an incandescent light bulb to match its brightness against the hot object as background. When the brightness fuses as one, the wire current is calibrated for known temperatures.

Convection Cycle

Convection is a fluid circulatory mechanism to remove heat. Circulatory heat removal is the most efficient method for low to medium temperatures. For extremely high temperatures and a very high heat loss rate is necessary, radiation qualifies better.

In a laboratory demonstration, a beaker of water is heated by a Bunsen flame from below. If tiny suspensions are in it as markers, a rising swirl will begin. Water directly above the flame rises in a column to the top, gets pushed aside to sink down the wall to fill the void left by the rising water. This is a convection cycle. In the UP flow, heat expands the water to lose density and rise obeying Archimedes's Principle. In the DOWN flow away from heat source, water cools by losing heat to its neighbour and recovers its original density. At the open top surface heat loss to air is greatest and the water cools fastest as well. This simple illustration of heat conversion into kinetic energy by water is the fundamental of climate physics but with a few amendments added by gravity effects over its greater heights and substitution of the Bunsen flame from below by the hot sunlight as heat source from above.

UP cycle. (SLOW *Inhale* of solar heat)
72% of Earth's surface is oceans, lakes, and rivers. Surface water is constantly evaporated into air by sunlight. Evaporated moisture harbours solar heat that it absorbs. Areas that receive more sunlight warm up faster to rise as moist air columns just like hot water in the beaker. Meteorologists call these air zones LOW because of its *continually** dropping air density by uninterrupted absorbing solar heat during the rise. Coriolis force twists this rising *warm moist* air counter-clockwise (CCW) north of the Equator.

*Note. Unlike the Bunsen flame that heats only the beaker bottom.

Fighting Earth gravity, this lifting by continuous solar heat is a slow tug of war in cooling the moist air through

1/ gravity extracting its internal heat as potential energy against height, and
2/ its expansion into thinning rarified air with rising altitudes.

The chill rate is – 3 C/ Km, or -2 F/ 1,000 feet. With ground temperature at a constant 15 C, at 5 Km above ground, moisture freezes into hexagonal micro ice crystals hovering as clouds. Having enveloped large pockets of air inside during *slow* condensation and freezing, these ice crystal are reflective white from high internal light reflection. Light reflection stops the solar heat inhaling process.

DOWN cycle (FAST *Exhale* of kinetic energy)
The sun disappears at dusk and the solar lift ceases. Unfrozen moisture cools by losing heat to surroundings, regains its high (1455 times) density on condensation to begin falling under gravity *with acceleration*. Meteorologists call these *cold falling* air zones HIGH because of their rising density. Coriolis force enters to twist them clockwise (CW) north of the Equator.

Completing this cycle, solar heat that is originally absorbed by moisture has been converted into kinetic energy inside falling water by gravity when the sun disappears. ***This is basically how Earth sheds ALL its absorbed heat everyday, NOT by radiat*ion.** However, this is NOT the whole story. There is a bigger story to be told about solar heat converts to kinetic energy without moisture at all in the upper atmosphere, where it is too cold for water to exist as a heat carrier. A specific type of *dry* air (+ ve ions) plays heat carrier instead. See Chapter 9.

Air from a HIGH never flows directly into a LOW. If they did, an abrupt violent storm would erupt. Instead their swirls connect with each other at edges of rotation gently to become surface wind in the same direction to avoid shears against each other. It can be seen that convection cycles of warm moist air UP (a LOW) and cold dry air DOWN (a HIGH) co-exist in Earth's atmosphere everywhere powered by sunlight *and gravity* to give us climates. There are three such *vertical* cycles on each side of the Equator as follows that depend on solar intensity;

1/ The strongest cycle occurs between Equator and 30 degrees N or S of the Equator
2/ The weakest cycle is between the poles and 60 degrees on either side of Equator, and
3/ An intermediate cycle between 30 and 60 degrees on each side of Equator.

It is necessary to have an odd number (3) of convection cycles to avoid vertical wind shears. Two rotations in the same direction must be separated by a rotation in between them in the opposite direction. If not, violent wind shears would generate storms by rapid condensation.

Humidity

While temperature measures atomic activity level, humidity measures wetness level of air. Except for very cold (way below freezing) N and S poles where all moisture is frozen out, even deserts retain a small residual humidity. Low humidity is the reason for desert hotness. Cooling down at night, moisture condenses as morning dew to sustain desert life like cacti, snakes and scorpions. At dawn, the rising sun quickly evaporates it back to air.

Warm humid air suppresses skin moisture evaporation blocking body heat loss as mugginess. In cold weather, high humidity in blowing wind makes it feel much colder (chill factor) due to extra heat loss from our body on contact to humidity that has a high specific heat.

Two important parameters are Relative Humidity (RH) and Dew Point (DP). For every temperature, pressure is from air molecules colliding with the wall. Presure depends on air density times the kinetic energy in air molecules. If moisture is injected into this air, the amount will increase until it reaches the same vapor pressure before it stops. At each temperature, dry air would tolerate a maximum moisture in equilibrium. This maximum is the moisture saturation point. The saturation temperature is the Dew Point (DP) for that moisture density. Any excess moisture would be rejected by condensing out as dew. This relationship of saturation moisture to temperature builds up a DP curve with temperature on one axis (abscissa on the horizontal) and saturation moisture on the other (coordinates on the vertical). A comparison between two points on the DP curve gives a Relative Humidity (RH) % of moisture density of the lower temperature point at the higher temperature.

The RH measuring instrument consists of two identical thermometers side by side. First one is called *dry bulb* which is a common mercury thermometer in open air. Second is a *wet bulb* with a wick wrapped around the mercury pool at its bottom and the free end dipped in a tiny pool of water. The wick draws water into evaporation for saturation to establish the Dew Point temperature. Dew Point moisture density divided by maximum possible moisture density at dry bulb temperature is the RH in percentage. RH shows how *close to* saturation the moisture content at any particular temperature is. Any temperature drop pushes RH beyond saturation to start condensation into rain or fog depending on the drop rate. Fog is from a fast drop. Slower drop allows droplets to coalescence into rain that falls to ground. In summary;

1. *Fast* **condensation expels heat from moisture in a rush as kinetic energy to spawn storms.**
2. *Slow* **condensation of moisture rising up to sky creates reflective white ice crystals as clouds that stop solar heat absorption.**

Under gravity, water seeks its lowest position for minimum potential. Moisture does the same. At first chance, gravity tries to split heat from moisture to return it from vapour to a liquid. On separation and condensation, the heat embedded cannot exit fast enough as heat but as kinetic energy in action. Exit rate decides the process. That is why a high RH is more prone to storms than low RH. Low RH delivers a clear blue sky (Rayleigh scattering/Tyndall effect) during day and a red glow in the evening (sailors delight). By contrast, low RH brings a red glow in the morning (sailors warning). Its thirst will make RH to climb fast by solar evaporation as the day wears on. Fast condensation brings storms. It takes only an abrupt temperature drop to flip it on.

Tornadoes

Should a *column* of cold dry air plunging down fast under gravity acceleration collide with a rising desert *hot* moist air, an abrupt *huge* temperature drop (-ΔT) spawns a tight twisting tornado over land or a waterspout over ocean instantly with freezing. Hailstones are flung out by the centrifugal force to smash down along with a characteristic fast churning tightly wound vortex that hovers drunkenly. So long as more falling cold dry air and rising warm moist air continue to fuel the collision, it continues. As soon as one or the other runs out the storm fades out suddenly, often in just minutes, and vanishes without a trace leaving only a trail of devastation behind. Tornado has a characteristic long tubular high speed churning vortex due to the rapid release of both latent heats of condensation and fusion (539 and 80 gm. calories/ gm respectively) together to yield hailstones. The plunging cold dry air column has overpowered the rising warm moist air. The wind vortex acts like a vacuum suction by Venturi Effect. The rising moist air must be very hot due to a large amount of solar heat packed into a low amount of moisture available for a short collision to bring about such a large temperature drop. The short lives of tornadoes can be accurately recorded to minutes in time that makes their occurrences similar to earthquakes and volcanic eruptions in suddenness and short span. It is a focused event. The short life of a tornado is due to the **dryness** of the rising hot air. Its low moisture content brings high temperature (a large ΔT) for a rapid chill but chokes the storm as fast as it occurs. The significance of that is revealed later in Chapter 13 under "concurrency". Waterspouts are rare because of difficulty in reaching very hot temperature over open water in the midst of a copious supply of moisture.

Hurricanes (Typhoons in Asia)

When warm moist air originating from a solar heated ocean rises up against a *layer* of dense cold dry air blanket hovering above in an inversion preventing its further ascent, a horizontal hot-cold interface is set up by the encounter to start a *slow* condensation. The temperature drop at the interface is small (-ΔT) compared to a tornado. Hurricanes do not generate high temperature due to the abundance of moisture to share the solar heat. Slow condensation takes days, even weeks, to develop into a massive system with a life of its own. Heat inside warm moist air *slowly* joins the counterclockwise (ccw) spiral as kinetic energy and a tropical storm is spawned. If the inversion layer fades or yields fast, the storm dissipates without any more condensation. If cold inversion layer is thick and persists to condense more rising warm moist air to fuel the confrontation, the tropical storm will grow bigger in size and in strength sufficiently to eventually either scramble or engulf the overhead cold inversion layer by flinging out water condensation centri*fugally* (outward) as rains. By then, tropical storm has come into its own as a full blown unstoppable hurricane with its angular momentum to further suck in more warm moist air as fuel by centri*petal* (inward) force as it hunts for further hot ocean water. A hurricane cannot stop until landfall where no more warm moist air is available for fuel. There its rotary kinetic energy is spent on impact, wreaking havoc. Unlike a tornado, hurricane vortex is just a thin wall surrounding a totally calm "eye". Storm surge, waves of water pushed in front of the wind does the major damage due to the high density of water.

Both storms demonstrate that solar **heat on inhale** is replaced **by kinetic energy on exhale** in a *fast* condensation that is triggered by a collision between a fast falling cold dry air under gravity acceleration with a slowly rising warm moist air lifted by sunlight. In each encounter, cold dry air comes out triumphant inevitably because cold dry air;

1. is denser, heavier with more momentum,
2. has gravity in its direction, and
3. is more powerful due to its origin (see AGBC in Chapter 9).

In the abrupt collision, cold dry air wrings out moisture in warm air like a wet towel by pushing it beyond saturation, past 100 % RH called the Dew Point (DP). The temperature drop ($-\Delta T$) in the collision condenses moisture into rain (if small T drop) or hail (if large T drop) and with gravity assistance expels internal heat as kinetic energy in wind as fast exit.

In any wind storm, the power source is at the hot-cold interface where **rapid condensation** expels heat from warm wet air to turn it into kinetic energy instantly to accelerate the wind. It is also where resultant rain (if only heat of evaporation is recovered) or hail (if melting heat of ice is added as well) are flung out by centri*fugal* force after they are formed. A centri*petal* force at the spinning axis that continuously sucks in warm wet air is what drives wind storm.

Hot wet air in a hurricane rises up from the warm ocean surface like a mountain to puncture through the cold dry inversion layer hovering above like a blanket. Where they collide is an upward pointing cone that is the hot-cold interface where heat exchange occurs and rain is flung down. Fastest wind in both hurricane and tornado is at its vortex in the spinning axis. The high velocity there makes it go sub-atmospheric in air pressure (< 760 torr or 14.7 psi) based on Bernoulli's theory of aerodynamics. In hurricane, wind is an **UP** ccw spiral against Earth gravity. As soon as hot-cold cone interface goes sub-atmospheric it sucks heavy cold dry air down from above to form a calm central cylindrical column called the "**eye**". Now the eye wall takes over as new hot-cold interface power centre with the fastest wind to continue its life. Rapid condensation within the Eye-Wall flings out cold rain by centrifugal forces. Where cold rain falls under gravity it forms a concentric curtain that creates a secondary hot-cold interface to repeat the process (called rain bands) that cascade out as concentric hot-cold curtains towards the storm edge. Each rain band turns heat into kinetic energy to augment the rotational angular momentum of the hurricane.

In a tornado, heavy cold dry air from above plunges down under gravity acceleration like a fast twisting drill bit into light hot wet air rising from ground below. Hot-cold interface is a narrow cone pointing **DOWN** in a ccw spiral. Once again the high wind velocity here makes it go sub-atmospheric in pressure. When it does this time, however, it is near ground level where surrounding warm air rushing in simply adds to its destruction power. No centre "**eye**" can be formed because the rushing-in warm air is lighter weight than the cold dry air in the central cone like smashing against a cold solid wall like stone. The in-rush warm air is drying fast as well having lost its limited moisture content on contact to initiating the tornado to begin with.

In both hurricanes and tornados, both centri**fugal** and centri**petal** forces are present simultaneously. The former discards condensation after its internal heat has been wrung out and converted to kinetic energy to power the storm, while the latter acting under Bernoulli cooling at the hot-cold interface at high velocity from the kinetic energy sustains a sub-atmospheric low pressure to continually suck-in more moisture, if any left, and sucks-down the fast spinning vortex towards ground level.

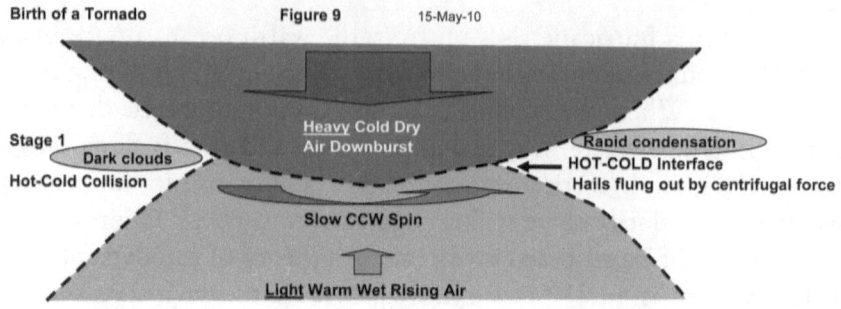

Birth of a Tornado   Figure 9   15-May-10

Stage 1
Dark clouds
Hot-Cold Collision

Heavy Cold Dry
Air Downburst

Rapid condensation
HOT-COLD Interface
Hails flung out by centrifugal force

Slow CCW Spin

Light Warm Wet Rising Air

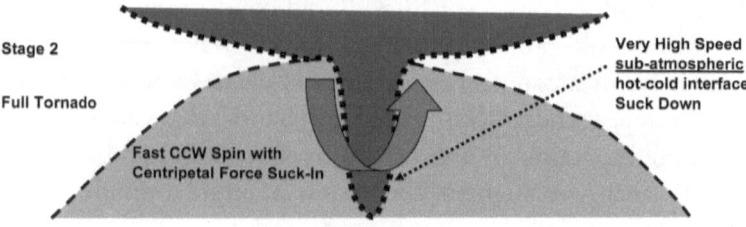

Stage 2

Full Tornado

Very High Speed
sub-atmospheric
hot-cold interface
Suck Down

Fast CCW Spin with
Centripetal Force Suck-In

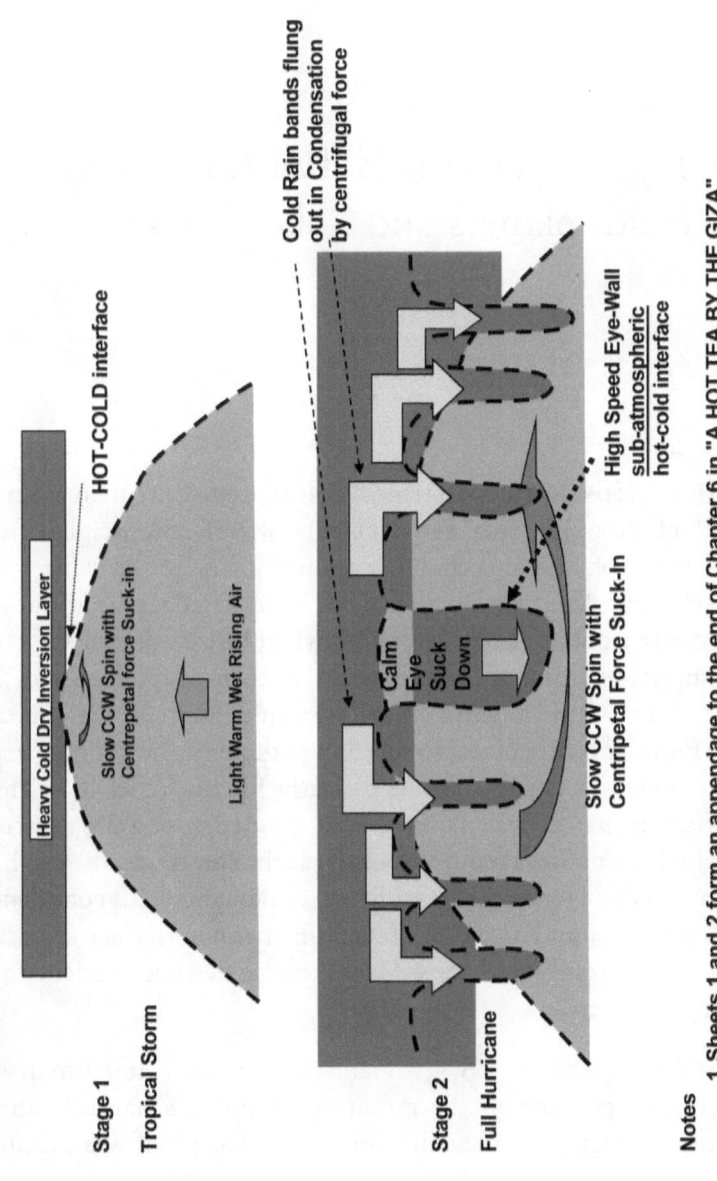

Birth of A Hurricane   Figure 8   15-May-10

Heavy Cold Dry Inversion Layer

HOT-COLD interface

Stage 1

Tropical Storm

Slow CCW Spin with Centrepetal force Suck-in

Light Warm Wet Rising Air

Cold Rain bands flung out in Condensation by centrifugal force

Calm Eye Suck Down

High Speed Eye-Wall sub-atmospheric hot-cold interface

Stage 2

Full Hurricane

Slow CCW Spin with Centripetal Force Suck-In

Notes

1 Sheets 1 and 2 form an appendage to the end of Chapter 6 in "A HOT TEA BY THE GIZA"

2 Cold rain bands of condensation that are flung out by centrigual force of the high speed "eye-wall" and falling under gravity (angle arrow) form concentric vertical curtains as new hot-cold interfaces to condense outer warm moist air cascading out radially to the edge. Each new hot-cold concentric curtain turns more heat into rotary kinetic energy to augment angular momentum of the storm

Chapter 7

Earth gravity, Coriolis Twist, Magnetic Field, planets and Solar Energy

Earth gravity

When Galileo (1564-1642) performed his alleged, but unconfirmed, experiment* of dropping two metal objects of different weights from the leaning tower of Pisa and they hit ground below at the same time, it was the very first demonstration of gravity and its effect. That did not get him into any trouble. Later on though when he as a devout Catholic threw his support behind a contemporary Nicolas Copernicus on his heliocentric theory (Sun at centre of universe) in defiance to the church's geocentric (Earth at centre of universe) edict at the time, he got nine years of imprisonment as a heretic. He paid dearly for his belief in science. Expert craftsman that he was, he improved the design of a 3X telescope to discover Jupiter's moon. Frequent looking at the sun tragically cost him blindness in a sacrifice for science. Galileo was a paragon of his confidence, servitude to discovery and purveyor of scientific truth in the face of hostile controversy and repressive authority. He was a true devotee to science and hero to many followers.

*Note. In 1971, Apollo 15 US astronauts David Scott and Jim Irwin confirmed the experiment on the moon by dropping a hammer and a falcon feather to reach ground at the same time. The proof was captured by a camera.

Galileo's death in 1642 was celebrated by the birth of his protégé Sir Isaac Newton, who was a genius in his own right. Later credited as father of gravity, Newton formulated gravity mathematically as an attraction force that decreases by the distance squared. Gravity not only governs celestial body motion in their orbits, it also sets up a framework of order for physical laws to work. Minus gravity, Archimedes would not have discovered buoyancy and his famous Principle. Chaos would prevail and objects would float around in uncontrolled fashion. Gravity gives context to Earth climate and in our quest to understanding climate change plays a big role to engender a *paradigm shift* towards possibly how global warming may be averted like coaxing the renegade legendary Genie back to his bottle and cork it. After all, gravity and its action on oceans has locked in another Genie under our feet for 4.6 billion years. More in Chapter 12.

Roughly 100 Km under our feet is a cauldron of swirling hot molten lava (a liquid plasma) mixed with hot gases that are eager to exit under tremendous pressure for 4.6 billion years. During its cooling down from a massive fireball, Earth's slow rotation of once in 24 hours pushed out a bulging Equator centrifugally on the outer crust at the expense of its N-S spinning axis. With diameter 7,926 miles (=13,210 Km) at the Equator, Earth is oval shape and 25.8 miles (=43 Km) fatter at its waist than its height (distance between the N-S poles). Its extra thickness adds 0.3% more gravity to an average acceleration of *9.8 meter/sec/sec* at the Equator. Equator is 41,500 Km or 24,900 miles at its girth. Correspondingly the atmosphere is also thicker over the Equator than over the poles by same amount. If dropped from identical heights at the same time, an object on the Equator would hit ground before its equal at the poles would.

Earth is over 72% under water. Pacific Ocean covers more than half that area. Water is lighter than hard rock. This imbalance of weight wobbles Earth in its spinning around the N-S axis. This wobble makes the magnetic N pole wander around its geographical axis.

Coriolis twist
Frenchman Gustave Coriolis (1835) observed that an object flying over different degrees of latitude on Earth would veer left (counterclockwise, ccw) or right (clockwise, cw) depending on direction due to Earth rotation. Rotation contains torsional (twisting) kinetic energy as in a flywheel. This torsional kinetic energy increases with radius from the axis of rotation. A man on top of Mt. Everest (8.848 Km) would be rotating faster but at same

angular velocity, (ω) than his friend at base camp below. He supplied that extra kinetic energy plus potential energy in his climb to higher ground. Unknowingly, anyone who travels from Toronto to Miami has exerted a similar effort as the Himalayan mountaineer. Coriolis twist is torsional energy to account for a latitudinal transition. It is why rocket launch picks locations closest to the Equator to utilize this bonus kinetic energy supplied by Earth's rotation.

Earth rotates slowly once in 24 hours at angular velocity (ω) of ($2\pi/24/60/60$) on an axis joining the N and S poles. Rotational velocity (V) is radius times angular velocity, ω. So, motion velocity depends on latitude. It is highest at the Equator where radius is maximum, and zero at the poles. To illustrate Coriolis twist, we use a Toronto couple in Canada going south to Miami for a winter vacation. Husband will fly his plane south overhead while wife is driving her car at exactly the same speed below. An imaginary straight road without curves exists and likewise for the flight path. They depart and arrive together by virtue of the same speed throughout the trip. To the wife in the car, her flying husband on takeoff would veer left towards Europe until he levels off at cruising altitude, then stays parallel overhead until descent to Miami when his plane will veer back to his right to land. After vacation and in returning home, the same will repeat itself, a left twist at plane take-off but right twist in final descent. One way to visualize this is to see that on leaving Toronto going south, the plane has to point ahead (left) on take-off to "catch up" to Miami that is rotating west-to-east faster than Toronto. On the return trip from Miami, the plane on take-off points backward (left) to slow down to match the slower west-to-east rotation in Toronto. North of the Equator, an UP motion (rising hot wet air) picks up a left (counterclockwise, ccw) twist while a DOWN motion (cold dry air) gets an opposite right (clockwise, cw) twist. North of the Equator, all UP motion incorporates a left-handed counterclockwise ccw Coriolis twist. DOWN motion incorporates a right-handed clockwise cw Coriolis twist.

Magnetic Field

Earth magnetism was first discovered by ancient Chinese over 4000 years ago, and has been essential for air, sea and land navigation ever since. Current literature attributes its origin to the combined action of earth rotation on molten lava circulating in Earth core setting up a so-called "dynamo effect". It does not explain WHY the lava flows in that way. I believe that I have a more plausible explanation here for the origin of Earth's magnetism. Nuclear fission of radioactive minerals like 92

Uranium 238 into daughter atoms is credited as the process for the 1050 C heat that turns rocks and minerals into magma, a liquid plasma, deep inside Earth's core. What is puzzling though is the ***endless*** source of energy to start molten lava swirling in convections that generate the "dynamo effect" for magnetic field. Even more bewildering is the spontaneity and abruptness with which molten lava can shift to bring about earthquakes and volcanic eruptions. What internal forces can cause all these events and without indications before hand? Might their origin be external ?

Unlike solid objects that rotate together in whole, fluids (both liquid and gas) only swirl in a spiral as seen in a tornado, waterspout, hurricane or a whirlpool because of a property called viscosity, or commonly called drag. The low viscosity of air permits car racing and airplanes flying at high speeds. Boats go much slower because of the higher viscosity of water. Air viscosity drops to lower density at higher altitudes to reduce drag to flights. That makes it attractive to jets to fly there to save fuel. With help from mountain ranges like Pamir, Hind Kush, Karakorum in the Himalayas of Asia, Alps in Europe, Rocky and Andes in America acting as stirrers, only bottom layer in the atmosphere (Troposphere) follows Earth's slow rotation at a snail pace of one rotation in 24 hours. The remainder upper atmosphere remains tranquil and stationary. For all intents and purposes, Earth's air is mostly ***fixed*** except for a thin 10 Km bottom layer near ground, out of a total 700 Km or 500 miles.

High above Earth between 80 Km to 200 Km is a thick but very rarified air layer that is a plasma of + ve charge O2 and N2 ions generated by powerful gamma (γ) rays in sunlight that strips the electrons off. Electrically, plasmas are like metals with excellent energy absorption and conduction properties. This is the Ionosphere that shrouds Earth like an orange peel. Its low air density and low viscosity does not follow Earth rotation below. Slow as Earth rotation may be at once in 24 hours, it is still a 21 to 52 KPH (= 5.8 to 14.4 meters per second) speed relative to a stationary Ionosphere over that long distance. This is very significant. If this plasma shield is sliced as tomato layers parallel to the Equator, each layer acts as a positive current of + ve ions moving clockwise (cw) against a counterclockwise (ccw) rotating Earth below if viewed above North pole from Polaris star. A charge moving in a closed loop sets up a N - S magnetic dipole along its central axis. Plasma ionic currents in each "sliced" layer add up into a magnetic field with the North pole pointing towards S pole. This magnetic dipole of Ionosphere would point in a N – S direction in reverse to Earth's own magnetic field. As shown in ***Figure 3***. The Ionosphere magnetic dipole would then induce an opposing N - S dipole on Earth.

Earth is a good electrical and heat conductor with its minerals and high water content as proven by each lightning strike finding a ground successfully. Molten lava is liquid plasma with electrons thermally stripped by the high heat. A liquid plasma subjected to a magnetic field would react ***inductively*** to create an opposing magnetic field. This is like a closed coil in which an induced current would flow to oppose an approaching magnet. I conclude from this that Earth's magnetic field is one induced by its rotation under the Ionosphere.

Earth's moon is the fifth largest ***solid*** object in the solar system after Earth, Venus, Mars and Mercury respectively. Its radius of 3,480 Km (2,160 miles) is just over ¼ of Earth's. Its gravity is known to be only 1/6 of Earth's. Surprisingly its magnetic field is hardly 1% of Earth's. Coincidentally, it has a negligible atmosphere. This correlation between having an atmosphere and its necessity for a magnetic field is unmistakable. This theory explains why Earth N-S rotational axis coincides closely with its magnetic axis. Magnetic N pole is known to wander in time slowly in response to adjustments made by the molten lava re-distributions due to the Earth wobble by imbalance in weight between land and water.

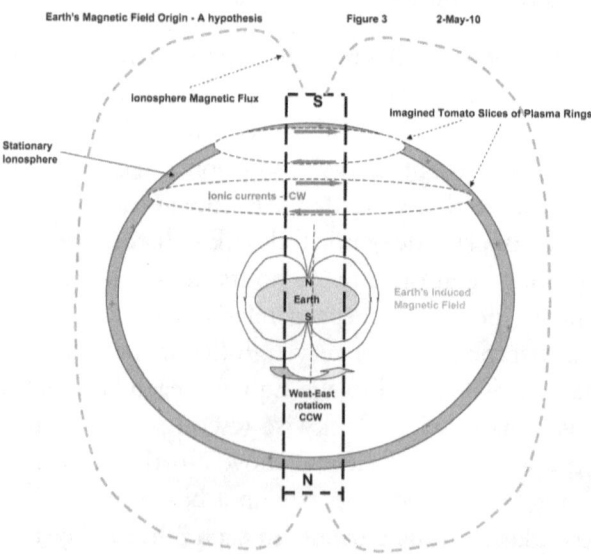

Figure 3. Excel graphic of Earth's magnetic field origin

Planets

With Pluto disqualified in 2008, there are only eight planets left in our Solar Milky Way. Earth is third closest to the Sun after Mercury and Venus. It spins CCW (looking down the North pole) slowly once in 24 hours but orbits CCW also even more slowly once in 365 Earth days around the Sun in an almost circular track. The two combined rotations plus a 23.4 degree tilt of the spinning axis relative to orbital plane provide four seasons with their weather. Inside the tropics between 23 N and 23 S latitudes from the Equator, the seasons vary little as the sun shines overhead each day. Further away though, summer is distinct from winter with spring and autumn in between. When the sun is north of the Equator from May to July, days in northern hemisphere are as long as 18 hours. Within the Arctic Circle (> 75 degrees N) the sun stays above the horizon at night as midnight sun. Conversely from November to January when the sun is south of the Equator, the Arctic Circle is dark even at daytime when the sun sits below the horizon. Location on Earth decides solar intensity and duration per season. This fact bears heavily on its local weather.

Averaged distance of Earth from the Sun (=1 AU, astronomical unit) is 93 million miles, a distance that takes 8 minutes for the speed of light, c = 300,000 Km/ sec. to cover. Earth receives sunlight 24/7 non-stop in its travel orbit, yet it manages to keep a rock stable *15 C* ground temperature for 4.6 billion years. Its atmosphere of 78% nitrogen plus 21 % oxygen and traces (0.03%) of carbon dioxide (CO_2) is distinctly different from its close neighbours on either side.

A neighbour orbiting at 0.7 AU from the sun is **Venus.** Its size and mass are similar to Earth but similarity ends there. Venus has a thick 99 % carbon dioxide CO_2 atmosphere. Venus is the *second brightest* star in the solar system only after our moon that is just 227,000 miles (=378,000 Km) away. Minus an atmosphere, moon temperature shoots over 100 C at noon but plunges to − 170 C in darkness. Venus is so near the Sun that she stays at a permanent stifling 460 C temperature, hot enough to melt tin and lead. This may have led to the notion of CO_2 as a heat trap that causes global warming. We shall debunk this wild assumption later in chapter 12.

Another neighbour on the far side at 1.5 AU from the Sun at is Mars. Its diameter is barely longer than radius of Earth. Her weaker gravity only keeps a thin atmosphere of 96% carbon dioxide CO_2 similar to Venus. Known as the *red* planet and notorious for its severe wind storms, Mars carries a dull orange glow. Farther away from the sun, Mars basks in an eternal chilling – 50 C low temperature. Two neighbours on opposite sides of Earth with similar CO_2 atmosphere but show an incredible 510 C difference in temperature ! So much for the CO_2 heat trapping theory in my view. What happens here ?

Solar energy

The Sun is a slowly spinning mass of very hot plasma of protons, H+ ions. Its surface is a hot corona at a temperature of 5000 C but at its core where two H+ ions are fused into a 2He4 helium ion (aka alpha particle) before being ejected exceeds 10 X*6 C. It outputs over 10 *22 KW of power continuously in radiation. This is a theoretical calculation as there is no possible way to actually do measurements. Solar energy obeys rules of radiation physics that lets demand control energy flow. Energy received by Earth depends on the coupling factor in resonance between sun, earth and conditions in between. To prove the point, a black surface next to a white surface absorbs solar heat faster by being a better coupler (impedance match). If earth coupling factor varies, then solar energy obtained will vary accordingly. This is the essence of radiation physics as so different from particle physics. Be very clear about it. Solar input is NOT a constant*. In an electrical simile, solar radiation is a force field like voltage (Volts) of a circuit but the current (Amperes) that defines energy flow depends on the **demand**. or load. It is a fact that afternoon sun is hotter than morning sun. The sun has not got hotter, nor the angle of incidence varied. The only increase is more water having been **evaporated** from lakes and rivers into the air raising its moisture content. Clearly level of moisture controls the coupling factor that sets the level of solar energy flow, as if Earth is darkening in complexion (tanning) with passage of time until dusk or the gathering of clouds to block further sunlight. ***Moisture is a capacitor equivalent in the solar RF circuit***.

*Note. Solar energy is an ultra-broadband of pulsating (E-M) force fields. Force field is like voltage (V) of a circuit that does NOT deliver energy. Once resonant coupling is made with a receiver, force fields are set into **motion** in a current flow (I) the value of which is based on circuit impedance (demand). If reflections from impedance mismatch occur along the way (including Albedo Effect) the input is accordingly adjusted, eg. "a silver lining" (edge) of a white cloud is brighter than sun.

Arriving on Earth, about 35% of solar energy is lost to the upper atmosphere, partly to an Ozone band up at 20 – 30 Km high and the Ionosphere between 80 – 200 Km, but mostly to moisture in Troposphere near ground. We see a blue sky (Tyndall effect) as a result of Rayleigh scattering of air molecules by blue light the most. Solar radiance on ground is close to 1 KW / square meter for convenient estimation. In temperate zones over 30 N or S far away from the Equator, sunbeam at dawn and dusk has to penetrate an air layer in excess of twice that at noon (zenith) and suffers attenuation. The intensity at noon on the Equator is defined as "air mass 1" (or AM1). Most solar intensity is quoted at AM1.5 being averaged between noon and twilight for design calculations. How sunbeam entering the atmosphere at lower angles and then affecting climate is dealt with in more detail in the following Chapter 8.

Photosynthesis is a process whereby chlorophyll incorporates green colour in sunlight and carbon from $CO2$ as building block for organic growth. It occurs both on land and in water. In water, aquatic microorganisms like planktons form the bottom of food chain. All organic end products represent solar energy storage. After millions of years under pressure and heat they become fossil fuels like coal, oil and natural gas. Fossil fuels are sunlight storage in time as carbon and hydrogen.

In terrestrial solar spectrum, half of its energy is heat near infra-red (IR) end of long waves. The remaining half in visible light (0.4 – 0.7 microns) to UV is capable of creating electricity *directly* by solar cells. Blue end of sunlight (shorter wavelength) is powerful enough to excite free electrons in silicon solar cells as electricity carriers. Longer waves beyond yellow colour cannot do it. The heat producing end does the heavy lifting of warming and cooling as shown in Chapter 10 to create a livable steady *15 C* ground temperature to sustain life on Earth.

Farther away from the Equator towards the N and S poles, solar daily variations follow the seasons closely. North of Equator, summer solstice (June 21) corresponds to the sun reaching the Tropic of Cancer (23 degrees N) and is about to reverse southbound. Winter solstice (December 21) is the opposite when sun reaches Tropic of Capricorn (23 degrees S) and is about to turn northbound. Midway between the two solstices are the two Equinoxes (northbound in March) and (southbound in September) when sun crosses the Equator. Earth decelerates approaching solstices and accelerates away from them. It does the reverse to Equinox, accelerates on approach and decelerates on departure. As a day on Earth is defined by one complete rotation on its N-S spin axis, Earth location in its orbit around the Sun in a specific season determines its daylight hours and locations on the horizon of daily sunrise or sunset. Variations are noticeably pronounced depending on if Earth is in acceleration or deceleration towards solstices or equinox in its orbit as its locations can change rapidly from day to day.

Standing on the Equator on a clear day without a compass, you can tell directions of East and West by sunrise and sunset, but not North or South. North of the Equator, the sun is to your south moving left to right (east to west) South of the Equator, the sun is to the north moving right to left.

NB. At publication time of the inaugural edition of this book, an idea related to designing and prototyping an ultimate solar power generator specifically for tropical regions has sprouted. After the usual patent protection and registration process in due course hopefully within a year or two and God willing it will be included in the next reprint as an additional Chapter 15 for general discussion at least if not in actual technical details. Conceptually this appliance can be mass reproducible at low costs everywhere by low to medium skill labour, and using indigenously available material to maximize self-sufficiency. It arose unexpectedly out of having written this book and having gained new insight as an extra bonus in the research for it. Writing can indeed be thought provoking !

Nae Ismail

Photovoltaic Systems

The manufacturer of photovoltaic cells is a smaller but rapidly growing segment of the Canadian industry. Photovoltaic systems are applied to specialized markets in Canada such as remote weather stations and navigation aids. A number of Canadian companies produce photovoltaic powered irrigation systems.

Les systèmes photovoltaïques

Les fabricants de cellules photovoltaïques, bien que moins importants, font sentir leur présence sur les marchés solaires. Les systèmes photovoltaïques sont utilisés dans des secteurs spécialisés du marché solaire canadien, par exemple, dans les stations météorologiques isolées et encore comme aide à la navigation. Plusieurs Sociétés canadiennes fabriquent des systèmes photovoltaïques pour l'irrigation.

Sistemas de fotovoltage

Los fabricantes de células de fotovoltage constituyen una pequeña, pero de rápido crecimiento, rama de nuestra industria solar canadiense. Los sistemas de fotovoltage requieren para su aplicación mercados especializados como una lejana estación meteorológica o auxiliares para la navegación. Un número pequeño de compañías se especializa en bombas de irrigación con energía de células de fotovoltage y sistemas.

1981

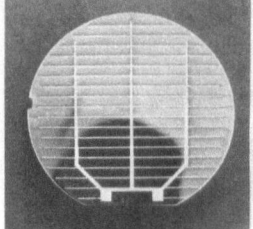

EnRel Energy SHS – 1000 Solar Home Power System

CHINA

TANZANIA - PRESIDENT J. NYERERE

HARARE, ZIMBABWE

KANPUR, INDIA

Chapter 8

Fluid Mechanics, Archimedes
Principle, Bernoulli Theory

Fluids

Under standard temperature and pressure (STP, 25 C and 760 mm Hg or 14.7 psi for one atmosphere), two out of three states in which matter can exist are fluids, namely liquid and gas. Fluids deform and flow in response to external forces (pressure) around them. The ability to flow allows fluids to make intimate contacts with solids or mingle with each other. Fluids are vital in energy transfer either from point to point, or as propulsive power in an expansion (phase change) from liquid to gas by injecting heat. Condensation (reverse phase change) from gas back to liquid is possible by squeezing out heat with a temperature rise if done slowly. If done abruptly, the rejected heat turns into a violent release of kinetic energy. Exit rate of heat determines which path to take. Fast heat exit is the cause of all storms.

Apart from the ability to flow, liquid and gas are different in *six* aspects as below;

	Liquid	Gas	
1/ Compressible	no	yes	
2/ Density gradient	no	negative	(by gravity)
3/ Pressure loss	linear	exponential	(by gravity)
4/ Viscosity	high	low	
5/ Molecular bond	weak to nil	no	
6/ Surface tension	yes	no	

Earth's atmosphere thins out over 500 miles (700 Km) to outer space. 50% of its mass lies within 11 Km above ground. At the top of Mt. Everest just under 9 Km, air is around 10 psi which is roughly 70% of its ground pressure of 14.7 psi, giving rise to breathing difficulties and risks of hypoxia to mountaineers. A fast dropping negative density gradient with height thus exists in the atmosphere due to air compression by gravity. This is how a helium or hot air balloon floats up gently and not pop up as a cork does in water. On its descent under gravity pull this negative density gradient cushions the fall to minimize chances of a hard crash landing.

In our laboratory demonstration of a convection current by boiling a beaker of water, it takes a while before convection begins. The time sets up thermally a negative water density gradient to create a lift. Scuba divers rise up *slowly* from the depth to avoid the much dreaded painful *"bend"* which would occur from a rapid release of dissolved nitrogen in their blood that was absorbed under high pressure during their descent earlier.

Archimedes Principle

When the famous Greek scholar Archimedes (287 – 212 BC) discovered the principle of buoyancy in his bath tub, shrieked "Eureka" and reportedly ran naked outside his house in euphoria, he had unknowingly changed the world forever. His discovery quadrupled human access on Earth from a land only low 28% to 100%, turned Indian Ocean into the most strategic water region in geopolitical military and commercial terms throughout all ages of civilization. It linked all manners of cultures between shores and opened up under water discoveries never imagined possible. His physics defines the law of how fluids interlace with one another by the effects of density under a gravity force. Simply stated, all fluids seek their lowest possible position under gravity. Heavier (higher density) fluids do it more powerfully than lighter ones to win the bottom layer. In other words, gravity tries to condense gas into liquid at first opportunity to raise its density in order to pull it down faster with acceleration. ***Gravity does the exact reverse of sun and its heat***.

Fluid Mechanics

Of the four possible states in which matter can exist, only solid is exempt from the Law of Fluid Mechanics. Liquids, gases and plasmas are fluids. In a fluid, intermolecular bonds of magnetic coupling have ruptured by thermal agitations beyond the Melting Point. Fluidity is from a pliant electron web that permeates between molecules and surrounds the periphery like a skin. Electrons are indifferent to heat. Except for water, liquids have a very low thermal expansion coefficient. At the Boiling Point, electron web that is related to surface tension has disintegrated and molecules break free individually as gas.

Four important parameters govern fluid behaviour. First is density (d) that obeys laws of gravity. Second is pressure (P) that sends energy for action. Third is temperature (T) that indicates density of kinetic energy per unit volume or activity level. Last but not least is velocity (V) if the fluid is in motion. For now, we would ignore (V) velocity for a dynamic case. We just focus on static cases.

Children play a game "rock-scissor-paper" to make decision. A rock can smash up a pair of scissors, which in turn cuts up a piece of paper. Paper can wrap up a rock. It shows the intricate dynamics of interlocking parts in a system. It holds true as well in a fluid system. Pressure pushes a fluid to flow. Pressure is due to gas molecular collision frequency on the containing wall multiplied by its molecular energy. Collision frequency depends on density (d) and temperature affects both molecular energy and density. Temperature has a swing effect on pressure by raising kinetic energy but dropping density at the same time. Solar heat is indistinguishable from human fuel burning. Heat is heat regardless of origin. All *variables* are constantly interacting with one another under temperature influence. Gravity is the only constant.

Heat drops fluid density to make it rise against gravity based on Archimedes Principle. Fluid density is monkey in the middle in a tug of war between solar heat and Earth gravity. In the day, solar overpowers gravity. Sundown, gravity wins in revenge. This is only true if working fluid remains a gas throughout. Should any condensation or evaporation, ie. a phase change in either direction, occurs in between, as moisture does in real life, a density quantum leap intervenes. *Dry* air is no problem as no phase change ever occurs. **Wet** air that is full of moisture is a different story altogether because of its definite phase change possibility.

There are *two* major differences between laboratory conditions and Earth atmosphere;

1. Sunshine comes from *above* like a microwave oven in space where temperature is a cold – 80 C. A gas being *continuously* heated is rising towards a colder condition.

2. Air being compressible, gravity has reduced air density with height by rarefaction. A solar warming thinning rising air struggles with a rarified cold thinner air while gaining and losing heat at the same time in a very complex dynamic. A battle ensues between a lifting hot sun against a gravity pull that makes it colder with height.

Moisture condensation triggers a density quantum leap of 1455 times and gravity grabs it *instantly.* Once it happens, all solar heat absorption slows down dramatically. In dry air, no condensation occurs. This is a first clue that *two* distinctly diverse climatic conditions exist in our atmosphere, one being moisture based and the other not. The effusive spreading nature of moisture dominates weather near ground where water exists in warm temperatures. Here, weather develops slowly in a track-able growth process. The other climate based on dry air *alone* can strike quickly over a random spot with its power focused. Dry air exists exclusively in upper atmosphere where moisture is frozen out by extreme cold. Moisture based climate is Tropospheric near ground only. Dry air climate is principally ionic in nature in the Iono-Stratospheric upper atmosphere with a different energy, which turns out to be more powerful and far reaching by its concentrated nature.

Helium (molecular weight 4) is 7 times lighter than air (molecular weight 28.8). Weighed down with sand bag ballasts that are discarded later in stages, a helium balloon rises as helium inside the balloon *expands* to find equilibrium until 35 – 40 Km above ground where any buoyancy left is cancelled out by payload weight under gravity pull. On the other hand, a hot air balloon rises under flame control to manipulate air density inside balloon. Hot air balloons reach only a maximum altitude of 20 Km. Air is so thin already there that further heating adds no more lift.

A full study of Fluid Mechanics is highly complex and beyond the scope of this book. Our interest is in understanding atmospheric events on Earth based on hot wet air and moisture contained therein. Moisture is the main heat carrier, therefore by default its climate agent, in the lower Troposphere by its ability to absorb heat and with a phase change property.

Bernoulli Theory

Swiss mathematician Daniel Bernoulli (1724) was credited as the theoretical father of flight before the Wright brothers (Orville and Wilbur) demonstrated in actuality their first powered flight at Kitty Hawk, N.

Carolina in the week before Christmas 1903. Observing that a fluid in high speed *laminar* flow loses internal pressure, he postulated how an object heavier than air might fly. His discovery equaled Archimedes's where an object heavier than water can float. Higher the velocity, greater is the pressure drop. Recalling that definition of pressure is molecular collision frequency times forces of collision on the containing wall, molecules in a fast flowing fluid direct most of their kinetic energy in the direction of flow, cutting transit time across the surface against which it flows low by its speed. Low transit time means less collision frequency with the surface. Low collision frequency times weaker collision force means reduced pressure against the surface. This is the Bernoulli Theory of flight, versus Archimedes Principle of buoyancy.

Bernoulli Theory provides lift to an airplane wing in flight. Adding a curve on top side of a wing speeds up air flow over it faster than air underneath. It creates a differential pressure between top and bottom to counter gravity while flying. Top and bottom air flows must both be *laminar* in layers without turbulence for it to work. Exhaustive air tunnel tests with smoke signals are routine to check scale models to gather test data for analysis.

One important side effect in Bernoulli Principle is in Cooling*. Pressure and Temperature drops are often synonymous during evaporation or (adiabatic) expansion in refrigeration cycles. Cooling at high altitudes above Earth where atmospheric pressure is low due to gravity demonstrates that fact. Pressure drop over a wing in flight chills the top as well. It poses grave danger if airflow over the wing contains moisture that could freeze into ice to adhere to the wing. Ice not only adds unnecessary weight, but ruins the critical foil curve for lift and laminar flow. This makes winter flying in cold temperature risky without de-icing the wings first by anti-freeze (which is ethylene glycol) spraying before take-off.

*Note. Some might argue it as Venturi cooling more correctly. I beg to disagree because no constriction orifice is used here that Venturi often needs to accelerate fluid to low pressure. As you will see, Bernoulli cooling uses density increase in condensation of the fluid and gravity for acceleration. Either way, it is using fluid acceleration to suck out heat. Normally it is the reverse application of injecting heat to produce fluid acceleration, as in an explosion.

Bernoulli cooling is **_not_** fluid circulatory cooling. Bernoulli cooling **_instantly_** sucks heat out of a fluid as kinetic energy via a **_fluid-to-fluid_** interface by a pressure gradient **_(ΔP)_**. Fluid circulatory cooling relies on a **_metal-to-fluid_** interface across which a thermal gradient (**_ΔT_**) exists and molecular collisions conduct heat away slowly. Blowing at a steaming hot drink before sipping it is Bernoulli fluid-to-fluid cooling, same as my Giza Bedouin host back in 1965 tossing a hot tea between two tin cups did. In Chapter 9, Bernoulli cooling is explained as a way for Earth to release hot plasma (**_dry_** ions) heat in Ionosphere without moisture nor condensation. Bernoulli cooling is how ALL planets without water expel heat.

Centrifugal and centripetal force

Centri**_fugal_** force has a radial component that points <u>outwards</u> away from the centre of rotation. Children experience this force by twirling a merry-go-round trying not to be flung off. Ancient warriors used this force in wars by swinging overhead a projectile wrapped in a leather pouch and on release hurled it at the enemy. Vehicle wheels exhibit this force in motion. Each point on a spinning body has same angular velocity (ω.). Tangential velocity (V) is rotation radius (R) times angular velocity (ω). Velocity (V) starts from 0 at the centre and is maximum at the edge. This implies an outward acceleration by an internal force pointing from the centre to the periphery. As rotation speeds up, centrifugal force will at one point rupture inter-molecular bonds to disintegrate the rotor. Centrifugal force exists only in spinning **_solid_** objects with molecular bonds, but not in fluids. Fluids would not contain a centrifugal force.

Centri**_petal_** force has a radial component that points <u>inwards</u> towards the centre of rotation. A swirling fluid is the only illustration for this widely misunderstood force. Hurricane, tornado, twister or waterspout fall in this category and their formation is explained by a centripetal force. The hurricane, or typhoon in Asia, is a SLOW builder by amassing warm (27C or 80F) moisture over a large swath of sun heated ocean surface as it tracks westward against Earth rotation. By contrast, a tornado is a FAST builder in an abrupt collision between hot 35 to 40 C moist air rising up from hot plains against a plunging cold column of dry air that has climbed over the Rocky pushed by a Pacific westerly onshore wind.

Newton's second law says a force produces acceleration. An object in front of a force moves faster than the one behind it. Under a centripetal force, a rotating fluid spirals into a fast spinning vortex that leaves the rotating plane to travel out along the axis of rotation pushed from behind by the mass. Fluid velocity is lowest at outer edge that is least congested with no compression, but highest at the vortex where its congestion-compression is maximum. Looking at two points within the swirling fluid at different radii, their velocity ratio is inverse to their radius ratio in order to conserve angular momentum, L = mVR. Centripetal force holds the swirling fluid together the same way a boa constrictor or python squeezes its prey to suffocate and crush all bones before devouring. Only fluids sustain a centripetal force, not solids.

It is a general mistake of science teachers to mimic centripetal force by twirling overhead an object on a string. The force in the string is a **tension** to oppose the centrifugal force generated by the object orbiting at its far end. A centripetal force produces acceleration whereas a tension in a string does not. There is a big difference.

Chapter 9

Heating, Cooling, Microburst or Avalanche Gravity Bernoulli Cooling (AGBC) Ionic Downburst

Heating

Heating is to raise temperature. Temperature indicates atomic activity level. Athletes know well the need of warming up muscles to deliver peak performance. Polar region motorists wait for car engine in cold winters to warm before driving off. Animals consume food as fuel (glucose) to burn with oxygen delivered by blood to generate body heat. Heat is life. We must maintain a stable 36.7 C (=98 F) body temperature to survive. If not, 40 C (104 F) is a killer fever, as is 35 C (95 F) the lethal hypothermia. Human life is fragile to within this tight **5 C (= 9F)** temperature range. We rely on plenty of cushioning in a turbulent heat world.

Heat energy transfer is done by contact or induction. Contact transfer can be conduction (fixed) or convection (circulatory). Induction transfer is by radiation. As will be shown later, induction is the only true heat transfer process in reality including even conduction.

Conduction

Science literature refers to molecules in a solid as particles in vibrations that depend on their energy content. Energy transfer is therefore by collisions of vibrating particles like billiard balls on a pool table. Molecules in a hot *solid* object would vibrate faster and with bigger amplitude than those in a

74

cold solid object. In contact together, faster particles would bump hard into slower particles to speed them up after each collision. Outside molecules of the hot object lose energy to outside molecules of the cold object, but their loss is replenished by molecules from the interior. Outside molecules of the cold object pick up extra energy and transfer it inward to those inside. A thermal gradient develops across the interface to flow heat from hot to cold, like water finding bottom in gravity.

Convection (circulatory)

If at least one of the two objects is fluid, a faster alternative energy transfer is possible by circulating the fluid in contact with the other to present fresh molecules continuously. In doing that, a dynamic moving interface replaces a passive static fixed one in conduction.

In theory, both the above sound plausible. In practice however it raises a spectre of doubt in my mind. Going back to Chapter 3 on Rutherford Paranoia, where molecules are 99.999 % empty space rather than solid particles, prospects of collision look bleak to move energy. The next section will show that a nucleus has both magnetic and frequency characteristics that allow it to couple with each other as radiation energy by resonance. Magnetism rises exponentially in close range. Conduction and convection interfaces (passive and dynamic energy transfer) bring molecules into close proximity for magnetic resonant coupling to take effect.

Radiation

Radiation energy must not be confused with radioactivity of the Marie and Pierre Curie variety. Unfortunately the same word is used for dissimilar activities. Radioactivity is nuclear fission of a heavy atom emitting *particles* when disintegrating into daughter atoms. Granted there are gamma rays (γ) and energetic beam emissions included, but they are adjunct radiations, and *not* particles. Radioactivity is emitting *particles*. Radiation is a pulsating electromagnetic (E-M) energy field *in motion* as energy flow that is invisible. Space is ink black. Astronauts do not see sunbeams outside spacecraft windows. They can only see objects that disperse sunlight, called Tyndall Effect. Light beam through a glass of water is invisible, but visible after a dye (very fine particles) is added. Same for the light beam in a movie theatre where dust or smoke renders the beam visible. Radiation mystifies many folks and is often misunderstood. A common misconception even by professionals is by treating radiation as a *particle beam.* Radiation is definitely NOT a particle beam*. If you fancied the sun as a light making factory shooting out light bullets (called photons) in all directions, stop IMMEDIATELY now. It is wrong. Your eyes deceive you. The example below shows how radiation works.

*Note. On May 21, 2010 A Japanese group launched an Akatsuki space probe with the "Ikaros" space yacht to explore Venus. The yacht is an ultra-lightweight plastic to test solar flight theory. It is expected to enter into Venus orbit by December. I can't wait to see if it works, or not.

Most towns or cities have a high water tower 100 ft high. Water is pumped into it until it reaches a pressure of 44 pounds per square inch (psi). One atmosphere is 14.7 psi = 32 ft. of water. So 96 feet is three atmospheres. Distribution pipes run into homes and buildings where water is used. Pressure is uniform throughout the system but until a tap is open, there is NO water flow anywhere. Water flows out an open tap at a delivery *rate* that is based on *demand.* In circuitry analogue, water pressure is voltage (V) and water flow is current (I) that is determined by circuit impedance (or resistance). Radiation field is like water pressure. Radiation is a very high frequency pulsating (E-M) force field that permeates 3D space but weakening with distance from the source, unlike the incompressible water where pressure stays uniform if there is no flow. There is no radiation energy flow until there is resonant coupling by a receiver, like opening a tap to begin water flow.

Being of high frequency, radiation obeys high frequency (f) rules that apply to capacitance and inductance in RF circuits. In a RF circuit, capacitance (C, $1/\omega C$) is the capacity of energy (E) transfer, and inductance (L, ωL) is the opposition to energy transfer. Together they limit the rate (current, I) of energy flow. Heat energy is radiation just below visible light at 0.8 micron down to longer wavelengths like microwaves. They can resonate and excite orbiting nuclei of conductors, such as water and metals. 2.56 Giga-Hz is frequency for microwave ovens to heat water. Food contains water and heating water cooks food.

Cooling

If there is heating, there must be cooling in order to sustain temperature equilibrium. Our 36.7 C body temperature is kept steady by a combination of urination, respiration and perspiration. These tasks are carried out to meet a cooling *rate* necessary by situation.

Solar heated moisture rising up through air cools *slowly* to condense and then freeze at last as white clouds. Slowness lets dry air trapped inside a water skin to expand and stretch by more heat to grow *big* bubbles. Big air bubbles have high light internal reflection to appear white to stop further solar heating. On freezing, these big bubbles become familiar white fluffy cotton-like floating hexagonal snowflakes that we see under a microscope. When cold dry air collides into a warm moist air, *rapid* condensation produces thicker water skin with less air trapped around smaller bubbles that look darker due to lower internal reflection. Higher water content is better in storing electric charges that originate from friction of collision. White clouds are dry and electric charge free. Cooling rate dictates how *warm moist* air condenses to different types of clouds. In a rapid chill without time for heat to exit, instant release of kinetic energy takes its place as a storm. Cooling is necessary for planets in the solar system to handle 24/7 solar heat while orbiting around the sun. Without cooling, they would burn up eventually.

In our solar system, only Earth is blessed with water and lots of it. Moisture is restricted to within the Troposphere that extends up to 11 Km high. Weather occurs by solar heat absorption (as inhale) then followed by kinetic energy release in condensation (as exhale) in Troposphere. Within this bottom layer of the atmosphere, moisture is heat carrier. Here near ground, heat release is water based using LTSC below as process.

The remainder of 500 miles (700 Km) upper atmosphere is only *dry* air and very cold at − 80 C. Moisture is excluded by freezing out. Two hot plasmas exist in this zone. First is Ozone band at 20 – 30 Km high. Second is Ionosphere at 80 – 200 Km high. Plasmas are very good solar heat absorbers like metals. Both plasmas must vent their heat somehow without help of moisture to transfer heat. Free + ve ions do that instead as explained below by HTFC.

Low Temperature Slow Cool (LTSC) – Tropospheric-Fluid Circulatory

Moisture inhales heat (solar or man made) to expand and rise by losing density. During its ascent* in a tug-of-war between solar heat against Earth's gravity, it teeters between being gaseous or condensing as water while cooling. At condensation, surface tension forms a water skin that traps air as bubbles. Liquefying at 4 C, its density quantum leaps 1455 times that it immediately falls under gravity pull. The only way for condensation to stay up high is to envelope enough thin air as large balloons. The balloon reflects and stops more solar heat absorption to freeze as ice crystals becoming white reflecting clouds. Heat inhaled is now as 100% potential energy high up at 5 Km. in Earth's gravity field.

*Note. Cooling rate with altitude is -3 C/Km (= -2 F/ 1000 feet). If ground temperature is usual 15 C, Mr. Everest peak at 9 Km. would be – 12 C (= -3 X 9 + 15) but gusty wind chill makes it feel more like – 20 C or lower.

On (heat) exhale, clouds and remnant moisture fall down as rain if it is warm enough to splash into ocean or lakes with kinetic energy to make waves, or hitting lawns and forests to disintegrate on impact. At the poles and on mountain tops that are cold, clouds float down as snow to retain some potential energy that is useful later (explained in Chapter 12). This process uses water as heat carrier. It is low speed, gentle, non-violent and routine every day without disastrous impacts to life. It is a fluid circulatory cooling process.

This slow cool dominates in summer months when an overhead sun at Zenith penetrates a thin atmosphere (Air Mass, AM1 scenario) to deliver 70 % of its energy to ocean or lakes to spawn hurricanes, tornadoes and other storms. The 30 % remainder of solar energy is lost to the Ionosphere and Ozone band in filtering out gamma rays (γ) and UV-B rays respectively.

High Temperature Fast Cool (HTFC) –
Ionospheric-gravity-Bernoulli

Above Stratosphere and higher is devoid of moisture and consequently immune to weather. Here is only rarified *dry* air. Between 20 – 30 Km is the Ozone band where lethal UV-B rays are filtered by O3+ ions. Without ozone filter, skin cancer incidents would be too much for life. Ozone band was noticed in 1980's by its frightening thinning (under 3 mm) at N and S poles from chlorinated fluorocarbons gas leaks that neutralize O3+ ions. That led to the ban of all CFC gas products except for Freon now only allowed in refrigerators. Between 80 and 200 Km sits the Ionosphere that filters out the even more lethal gamma rays (γ) in sunlight. This zone was famous for blocking out spacecraft telecom with Earth on re-entry. Both Ionosphere and Ozone band are active plasmas that absorb solar heat to reach 2000 – 5000 C high temperature. Without water, how do they cool?

This cooling dominates in winter months in polar and sub-polar regions (over 45 degrees latitudes on both sides of Equator) where a low angle oblique sun skimming along the horizon delivers most (perhaps 65 %) of energy to penetrate a thicker upper atmosphere in an (air mass) AM 4 to 6 scenario, leaving little (around 35 %) portion to reach ground. This large amount of heat in the upper atmosphere powers a very strong cold Jet Stream. In my view, winter cold is not just from a weak sun. It is *made* cold by sunlight pulling down – 80 C cold air from space. A low angle horizon skimming sunbeam penetrating a long path in the upper atmosphere actively brings chill throughout winter. There is zero doubt to me that this happens on Mars to power her ferocious red dust storms that until now have defied an explanation. Mars has no detected atmospheric moisture but yet maintains a steady – 50 C cold temperature ! How else can Mars cool ?

Microburst or Avalanche Gravity Bernoulli Cooling (AGBC) Ionic Downburst

Since the Wright Brothers (Orville and Wilbur) demonstrated their first powered flight at Kitty Hawk, North Carolina the week before Christmas 1903, there has been periodic reports of airplanes having vanished mysteriously and suddenly in *mere seconds* without distress calls in the infamous Bermuda Triangle between Bahamas, Miami and Bermuda of mid-Atlantic Ocean. Pilots who survived quoted onboard navigational gears and compass going berserk and unresponsive as a rule. This area is among the busiest maritime and aviation traffic in the world. Any type of disappearance is quickly noticed. There was of course the Lockheed Electra piloted by a courageous heiress pioneer aviatress Amelia Earhart on her epic historic around-the-globe attempt that tragically fell out of the sky (July 2, 1937) in Pacific Ocean. Since then for over two centuries there was a growing list of inexplicable high profile aviation incidents including Swissair 111 (1998), Egypt Air 990 (1999), balloonist Steve Fossett (2007), Yemenia Airbus 310 (2009) and Air France 447 (2009) plus many others unreported due to without witness at the scene.

Meteorologists explain these accidents by "Microburst". In theory, they claim that rain falling from clouds on meeting dry air would quickly evaporate that chills the air. Colder air descends faster by gravity acceleration (that is true) to create a microburst. There are *five* serious problems with this theory.

1. Dry air over a vast ocean is impossible. Dry air exists only above the Troposphere and over the N and S poles where extreme cold freezes out all residual moisture. Even deserts retain a tiny moisture content that absorbs solar heat to make it hot. Without any moisture, sand reflects light and stays cool in fact.
2. A falling rain in broad daylight is unlikely because moisture absorbs solar heat to keep it rising up by expanding, instead of condensing to fall.
3. Rapid evaporation to bring such chilling can only happen with extremely fine mist, like fog, that is thoroughly mixed in air and certainly not from rain drops. Such fine mist does not fall, but hovers as water is lighter than air. Wet air rises.
4. Surrounding dry air, if it exists, on cooling by evaporating rain as proposed gets wet to rise even faster, and not to fall as a microburst. Finally,
5. Microburst cannot explain electronic interference on instruments and compass.

A more scientifically sensible explanation is hypothesized below not only for violent abrupt downdrafts to flush airplanes out of the sky and fouling up their instruments, but also for Jet Streams and ferocious red dust storms on planet Mars that are visible by a simple telescope. In truth, it also has close ties to earthquakes and volcano eruptions.

The Stefan-Boltzmann Radiation Law states that radiation loss is a function to fourth power* (***Figure 4***) of its absolute temperature. Its ***rate of change*** (tangent to the curve) is therefore the first derivative by temperature in Isaac Newton's differential calculus. So, radiation loss rate ***change*** would be a third power function in absolute temperature with 4 as coefficient. As an example, heat loss ***rate change*** (gradient of tangent to radiation loss curve) at 1200 Kelvin is 7** times faster than that at 1000 Kelvin degrees, only 200 degrees less. Solar heat absorption is linear with time, and so temperature must rise linearly as well. Radiation loss climbs faster non-linearly to fight warming. Imagine for every ***one*** degree climbed above the previous high mark on a temperature ladder, radiation loss slips back 5 degrees at first, then 20, or 100, and even 500 degrees for each degree climbed higher. It gets steeper, less stable and more slippery the higher it goes. At some ***critical step of no return*** the slip-back becomes unstoppable into an avalanche. Analogy is a Himalayan mountaineer initially falling down near base camp, but recovers to climb higher. He then falls again at higher altitudes but still recovers to go back climbing until he falls at the vertical notorious Khumbu Ice Ridge on Mt.

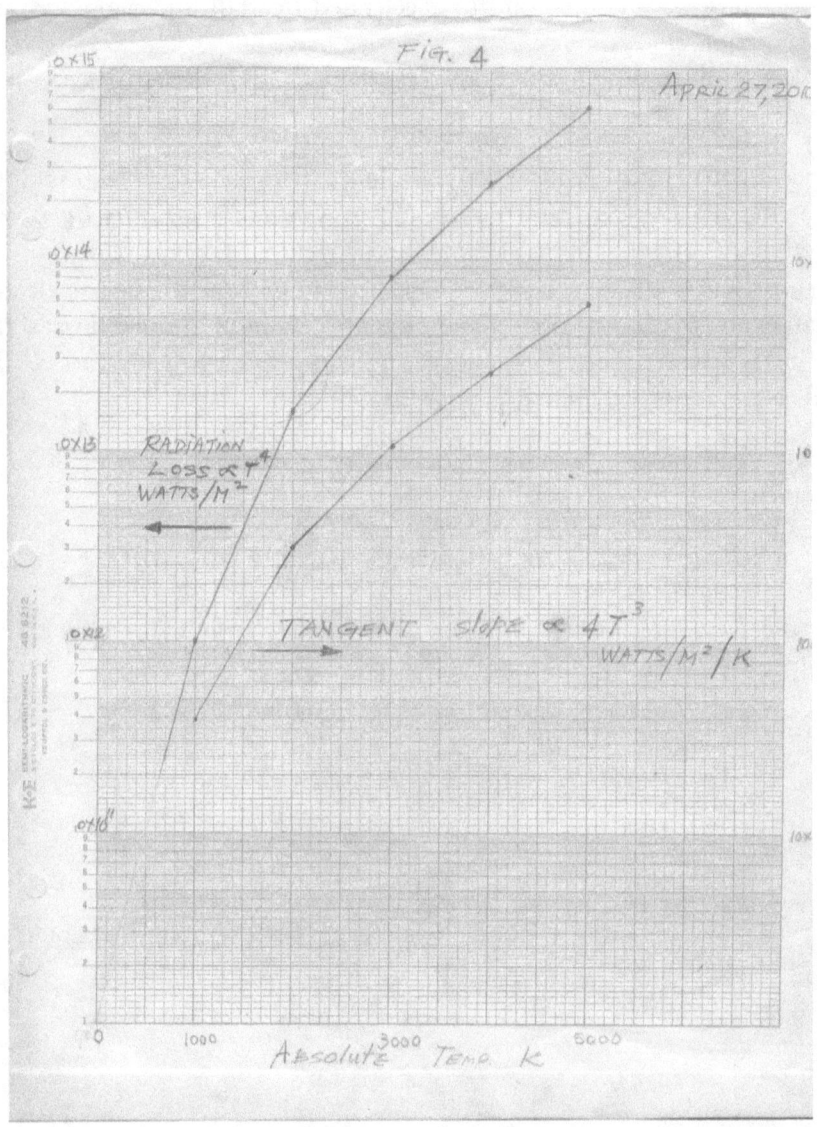

Figure 4. Graph of Radiation Loss vs K (Absolute temperature) and

Everest that has killed many climbers. One slip, the fall is non-stop. In a gas, this has important implication to its density rise during a plunge. A plasma *slowly* heating up by sunlight reaches a ***critical*** temperature whereby; (***Figure 5***)

(1) an avalanche massive radiation heat loss occurs abruptly. A massive heat loss triggers a

(2) sharp density jump to begin accelerated free fall under gravity pull. Acceleration creates

(3) Bernoulli cooling in fluid flow rapidly to raise density further to start a positive feedback loop

Once it enters into a positive feedback loop, there is no reversal (braking). A jarring 1-2-3 tandem sequence creates an ***Avalanche Gravity Bernoulli Cooing*** (AGBC) of heat into a rapid fluid flow kinetic energy of dry air ions ***instantly.*** A hot plasma becomes a gushing ***ionic cataract*** of + ve electric charges at – 80 C. Any poor planes that are caught in its downburst stand no chance with all electronic gears and compass electrically messed up. This is the most convincing hypothesis.

*Note. The radiation laws says that radiation loss, $R = \int T*4$, that is a steep curve of T

So, rate of change to radiation loss (tangent to curve) $dR/dT = 4T*3$

**Note. 7= 4 X (1200/1000)*3. A 700% jump in heat loss rate for a 20% temperature rise.

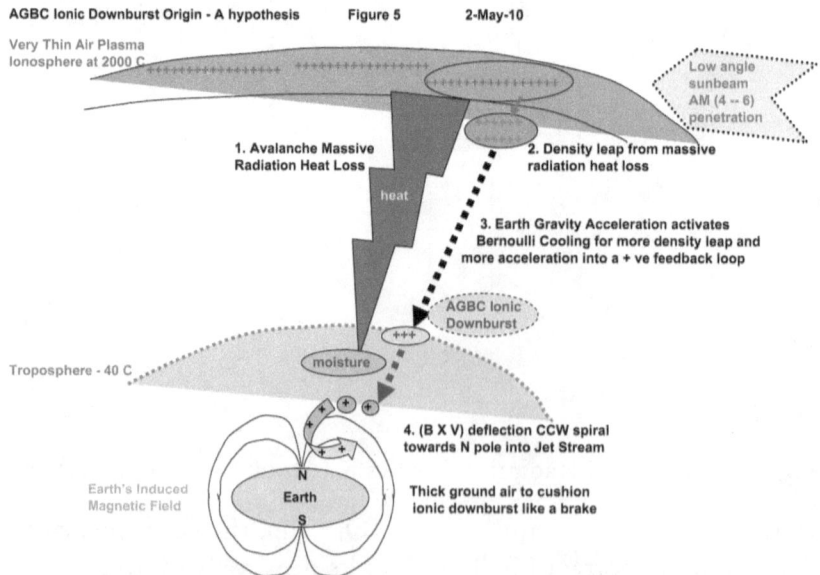

Figure 5. Excel graphic of "AGBC ionic downburst formation and Jet Stream

For the fun of it, we can estimate how fast an ionic downdraft can be.

By energy conservation law,
(kinetic energy) $\frac{1}{2}mV^{*2}$ = mgh (potential energy)   (1)
We determine velocity, $V = \sqrt{2gh}$ (2)

An object free falling without air resistance from the **Ozone band** at 20 Km high would reach a velocity of V= $\sqrt{2}$ X 9.8 X 20,000 = 198 meters/ second or 712 KPH

A condensing plasma (fluid not solid object) plunging down in air even under gravity pull may be slowed by a factor of 10 to reach velocity of 20 meters/ second or 71 KPH. This is still plenty downdraft to flush a plane out of the sky.

Looking at the Ionosphere plasma, an object free falling from 80 to 15 Km (height of Jet Stream) would reach velocity, V= $\sqrt{2}$ X 9.8 X 65,000 = 1129 meter/second = 4,064 KPH.

Discounting by a factor of 8 for a condensing plasma entering a thinner sub-atmospheric air, A Jet Stream (without accounting for spiraling around magnetic field flux) would get a speed in the range of 141 meters/ second = 508 KPH.

This abrupt heat-to-motion delivery of a **cold ionic** downburst can happen anytime, even in bright sun unlike the slow circulatory cooling by water in the lower Troposphere. I submit that ferocious red dust storms on Mars cool down the planet that way. The same process on Earth creates Jet Streams in Stratosphere with some help of Coriolis force in making + ve ions spiraling with velocity (V) (counterclockwise, ccw in the North) down Earth's magnetic field flux (B) under a (BXV) deflection torque.

In summary, planets without **_water_** cool by triggering;

(1) an avalanche radiation massive heat loss followed by a
(2) corresponding massive density jump, that subjects it to
(3) instant gravity acceleration and finally
(4) Bernoulli cooling in tandem into an irreversible positive feedback loop.

The initial avalanche heat loss is via resonant coupling to moisture in the Troposphere below that is eager and ready for any heat flux. Indirectly, moisture still plays a part in upper air (+ ve ions really) cooling.

For a real life simulation of this abrupt event of gravity pull, flush a toilet and watch water in the bowl disappear in a loud wh-oo-sh or gurgle. Water released by the tank raises bowl water level over a gooseneck behind the bowl, and gravity enters to suck out the entire lot. Abruptness of its start, intensity of energy flow, and the path it follows is like a lightning bolt. It is over in fractions of a second, leaving a wake in its trail. It also resembles a laser action where electrons are optically pumped up to an excited state waiting for a trigger to fall back together at once to emit a pulse that is coherent both in space and time, or phase locked in complete unison as massive power pulse. An AGBC ionic downburst is very powerful in its destruction power that could rival a lightning bolt in speed and trajectory.

Chapter 10

Earth Atmosphere, Fog, Corona and Jet Stream

Earth atmosphere

Earth's gravity of 9.8 meters/sec/sec. wraps us in an air blanket 700 Km (500 miles) thick, roughly 12% of Earth's radius of 3963 miles. First 2 miles (0.5 %) control 99.9 % of our existence. We exist on Earth like parasites on our skin. First 8 miles (2%) contain 99.9 % of moisture that drives the weather. Above that, there is only rarefied calm steady dry air and plasmas (positive + ve charge air ions.) which is hostile and un-inhabitable but however not totally useless to us.

Our closest celestial neighbours Venus and Mars have similar atmospheres but with carbon dioxide (CO_2) instead of air. Air is oxygen and nitrogen in a 78 to 21 % ratio. Mysteriously Earth keeps a rock stable 15 C ground temperature despite 24/7 sunlight bombardment. So far, Earth is the *only* planet known to have life in our entire universe.

Troposphere

This comparatively *thick* air layer extends from ground up to 11 Km high, carrying 50% of its total weight, but 99.99 % of its humidity. White reflective clouds hover at its ceiling it to block sunlight from being absorbed below. There is a steep negative temperature gradient of – 3 C/ Km (or -2 F/ 1000 feet) throughout this air layer by virtue of combination of two effects as follows; (*Figure 6*)

1/ adiabatic (fixed energy content) expansion when rising wet air enters low density zones at high altitude,
2/ internal heat of wet air is extracted as potential energy by Earth gravity.

Its ceiling cools to a low – 40 C as a result. Closest to ground, Troposphere is the only air layer that swirls with Earth's slow rotation of once in 24 hours. Rest of upper atmosphere stays relatively still due to a fast dropping viscosity (drag) with height. Earth surface is 72 % under water leaving less than 28 % as land. On land, there are lakes, ponds, rivers and vegetation that contain water. Thankfully, Earth is extremely water rich except for dry deserts. Polar ice caps and hilltop glaciers, although water as such, are also **dry because** moisture is frozen out by cold. Moisture is perfect for heat absorption as each individual molecule (a very unique electric +/- dipole) responds well to all modes of excitation.

Gravity engenders a heavy settlement of invisible microscopic dusts, pollens, spores and other ultra-fine particles at ground level and above open water. They are nucleation centres to which moisture molecules can cling after solar evaporation. The unique +/- electric dipole of moisture molecule attracts one another to lattice up like magnets into a web around nucleation centres. Moisture is an excellent heat absorber. In daylight, solar

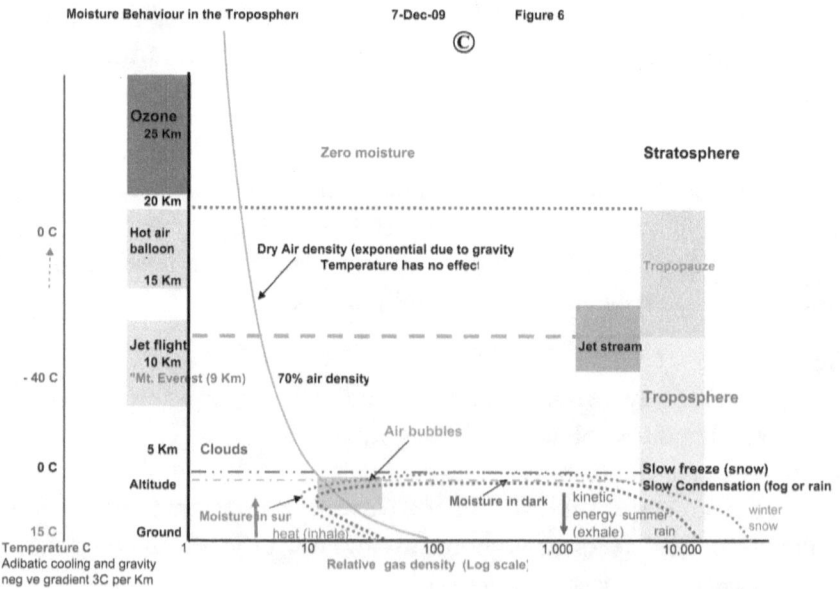

Figure 6. Excel graphic of "Moisture in the Troposphere."

heated moisture molecules expand to rise up. During ascent, air trapped in the nucleation centres expands to offer more surface areas for more water molecules to lattice up. These moisture clusters grow bigger as they rise like clear balloons. Moisture rising against gravity also chills by the processes cited above in a constant tug-of-war favouring the overwhelming sun. At condensation around 4 C, microscopic frozen bubbles are formed. This UP process is tantamount to inhaling solar heat and stores it in white reflective clouds (ice crystal bubbles) in the sky as potential energy. Heating stops when white clouds reflect sunlight

Stratosphere

This air layer starts above the Troposphere and extends to 50 Km above ground with air thinning out exponentially with altitude. At its bottom (20 – 30 Km) is a highly active Ozone (O3+) layer that filters out lethal UV-B rays in sunlight. Ozone layer is a plasma (free mobile ions) created by oxygen atoms losing some electrons to UV-B rays. A plasma of *heavy* (MW 48) mobile positive + ve ions is a good absorber and conductor of heat and electricity just like metals. Ozone layer absorbs sunlight and reverses the temperature gradient from negative in Troposphere to positive here to reach 0 C at its ceiling during in sunlight. This solar heated plasma is power source of AGBC ionic downbursts.

There flows a west-to-east Jet Stream of hundreds of Kms wide at speeds to 500 KPH and at the demarcation between Stratosphere (16 Km) and Troposphere (7 Km) near N and S polar regions. Pilots take advantage of this free ride to save time and fuel. Jet streams and Ozone band in Stratosphere are strategic players in Earth's climate.

Mesosphere

This layer starts above the Stratosphere (50 Km) and ends around 80 – 100 Km above ground. Little happens here apart from separating active layers above and below it. It reverts back to a negative temperature gradient by same adiabatic expansion which extracts heat for potential energy ending up with an extreme cold of -110 C at its ceiling.

Ionosphere

This is the outermost air layer, from 80 Km to 200 Km above ground, like a skin to Earth. Air here is so rarified that its air molecules are collision-free. Freedom from collisions permits each air (either O2 or N2) atom to accelerate to extreme velocity but still under escape velocity of 5 – 11 Km

/second = 3 - 7 miles/ second, Mach 34 times speed of sound to remain with Earth. High velocity means high kinetic energy (E= ½mV*2) and high temperature of 2000 C to 5000 C. It would nonetheless still "feel" cold (-140 C) to touch because of extremely low air density. Its high energy comes from absorbing the most powerful 10 – 100 MeV gamma rays (γ) rays in sunlight that are 1000 times stronger than 10 - 100 KeV X-rays. Ions of such velocity are able to Corona (faint glow) in the UV spectrum.

Ionospheric plasma as an electrical shield during space mission re-entry and as a wall to bounce HAM and CB radio signals is well known. Its role in climate control is yet poorly understood. Its role on Earth's magnetic field was hardly investigated either. My own familiarity with solar energy and plasma physics expect that Earth's magnetic field has to co-exist in harmony with the Ionosphere, and not *in spite of* it. That led me to believe that Earth's magnetic field is an induced one (opposing) in response to its rotation relative to the stationary Ionosphere above. See Chapter 7 argument.

As plasma, the Ionosphere absorbs solar energy gamma rays, (γ) like metals could. For polar and sub-polar regions, a summer sun is zenith high that has a short transit time going through the plasma. In winter when sun switches to low angle of incidence gleaming along the horizon, it relishes a much longer transit time penetrating the plasma. Longer transit time in winter delivers a far greater amount of heat absorbed by this plasma contrary to conventional wisdom. Upon release on trigger (see AGBC in Chapter 9) this heat storage avalanches down in an abrupt cold *ionic* downburst plunging to ground to spawn storms of greatest violence. When this ionic cataract crosses Earth's magnetic fluxes, a (BXV) deflection torque steers it spiraling towards both N and S pole to form those mighty Jet Streams. It is also how both N and S poles develop their severe cold aggressively, not just from a passive lack of sunlight.

Its sheer enormity and overwhelming presence totally envelops planet Earth like a mother's womb. Only her integrity and stability can guarantee us a stable and safe life. Its *gradual and smooth* transitions while Earth orbits around the Sun give us four regular seasons. Any untoward disruptions to the Ionosphere (*especially those intrusive jet and rocket flights that dump heat, turbulence and nasty exhaust fumes into such a pristine environment*) that invoke contortions here

could cause Earth's magnetic field to shift in *abrupt* rearrangement of lava flow in Earth's centre core as compensation. Lava flow re-configuration acts on tectonic plates and tips the delicate balance among them. Outcome could be either a volcanic eruption or an earthquake. This illustrates the holistic connection* between events high up in the attic to a reaction below in cellar of an interlocking magnetically coupled system*.

*Note. Our knowledge of the mysterious Ionosphere is still very embryonic. On March 31, 2010 media reported of a publication in Journal of Zoology by British biologist Dr. Rachel Grant of her 2009 observation in Italy of a common toad (Bufo bufo) that had vanished massively 10 days before a 6.3 earthquake at L'Aquilla that happened on April 6, 2009. Later checking with Russian data of electrical activity in the Ionosphere by radio receivers corroborated with VLF (very low frequency) energy bursts on the disappearance date of those toads. Surprise ?

Fog

Fog is a mass of microscopic water droplets suspended in air that is condensation out of saturation in a gentle cooling but before freezing. Slowness allows heat release without violence. Moisture is lighter than air by a 18 to 28.8 density ratio. On condensation, water density jumps by 1455 folds to fall as rain if coalescence occurs. Fog sits in a delicate balance between moisture and air found often at junctions between warm and cold ocean currents, like the Grand Banks of Newfoundland in North Atlantic.

Early morning fog before sunrise is also common over calm lakes, ponds and valley hillsides after gravity has time all night performing a delicate task of stripping wet air off its absorbed heat from the previous day and to settle down at its lowest point. The same process creates morning dew on our lawn, car windows, and in deserts. It is a big part of the Earth cooling (DOWN, exhale) process each night to eject heat before the next daily solar cycle resumes.

Corona

Corona is a plasma that emits sustained radiations. Industrially, corona is employed to "sputter" coat very thin target material to a surface to impart special properties. Target material atoms are forcefully ejected by energetic heavy ionic bombardment in vacuum to coat the surface like spray painting. A heavy gas (argon, M.W. 40) with a cooling gas (helium, M.W.4) is mixed in a high vacuum (low pressure of 100 milli-Torr, mm Hg) between the target and the coating surface while radio frequency (RF) power is applied to the two objects acting as electrodes. The visible faint glow is a corona where heavy argon + ve ions are accelerated to high velocity smashing into the target. A thin dark space separates the corona from the surface to be coated. That is a zone of electron clouds collected there by their light weight and high mobility compared to the heavy argon + ve ions. An Earth corona would be hard to see by naked eyes in daylight, especially if it falls in the invisible UV spectrum. I believe that ***distant*** planets that glow bright at night possess an outer corona of atmospheric CO_2 + ve ions (M.W. 44, heavier than argon) and not just sunlight reflection like our moon that has no atmosphere and is much closer to see. Moon is the next brightest solid object in solar system, after the sun.

Jet stream

Jet Streams are high speed winds that flow exclusively in polar and sub-polar zones at the interface between Troposphere and Stratosphere in a west-to-east direction. They can be several Km thick and hundreds of Km wide. Its origin has puzzled meteorologists. Jet streams play only minor roles in daily local weathers but exert enormous **seasonal** influence on regional climatic patterns because of the tremendous energy that they carry.

In the course of a day, sunlight spends more time in upper atmosphere than in the lower Troposphere. In a 12 - hour day, solar energy touches ground from overhead between 10 AM to 2 PM for 4 hours through a thin atmosphere (air mass AM1.0). The rest of 8 hours sunbeam shines at low angles close to horizon (air mass AM 4 to 6) penetrating the upper atmosphere. More time spent in plasma layers leads to greater source of heat energy in Jet Streams. Jet streams have their origin in solar energized positive charge + ve air ions in the Ionosphere.

+ ve or − ve charge particles cutting across Earth's magnetic field (B) with velocity (V) are deflected into helical spirals (in opposite directions of course) around magnetic flux lines to the nearest pole (upstream) by a (B X V) torque. High velocity (V) beta rays (- ve electrons) from sunbursts upon entering Earth's magnetic field (B) spiral right-handed clockwise (cw) towards N pole in summer months to create spectacular Northern Lights by ionizing air in their paths. Hot plasma + ve air (O and N) ***ions*** crossing Earth's magnetic flux lines spiral ***left-handed*** counterclockwise (ccw) towards N pole , but right-handed clockwise towards S pole south of the Equator. A left-handed counterclockwise (ccw) spiral towards N pole creates east-to- west Jet Streams. A right-handed clockwise (cw) spiral towards S pole also creates a west-to-east Jet Streams at the S Pole.

Jet Streams are stronger in winter when sunbeam comes in at its lowest angle from the far end of opposite hemisphere when penetration through Ionosphere is the thickest. Cold Jet Streams create frozen Arctica and Antarctica. The latter is much bigger in area than the former for good reasons. There is much more open water than land at the S pole. In a southern summer, there is feverish ocean evaporation around the S pole that turns into tons of snow when southern winter arrives. Rich moisture accounts for extreme ferocity and unpredictability of notorious storms to mariners sailing the South seas. No rain ever fell in Antarctica. Extreme cold freezes out any residual moisture in air, making it bone dry and heavy. Weak sunshine alone cannot produce extreme cold at both N and S poles. Extreme – 100 C cold from outer apace is piped in by high speed dry air ions of the Ionosphere powered by sunlight from the opposite hemisphere.

In comparison, N pole is surrounded by more land than ocean. Less ocean evaporation in a northern summer condenses into a small Arctica in winter. Higher heat retention of land also delays the onset of winter at N pole, which nowadays is further exacerbated by heat of human activities that is also more heavily clustered on both coasts of N. Atlantic Ocean. On top of all that is the famous Gulf Stream that brings warm Equatorial ocean current up the east coast of N. America to the North Sea. As a result, global warming would be more noticeably pronounced in the north than south. It demonstrated well beyond any doubt in the winter of 2009-2010 by the cascade of unexpected blizzards, tall ship sinking, torrential rains, quakes and volcano eruptions, mostly north of the Equator.

Chapter 11

CO2 greenhouse gas (GHG) controversy arguments, and Solar duvets (2)

Greenhouse effect was first reported by Joseph Fourier in 1824. Placing a glass cover over a box in sunlight in a crude experiment, he reported a temperature rise inside with time. **He had no idea if any moisture was inside the box**. He argued that visible light grew longer wavelengths (heat) after passing through the glass that were later reflected and trapped as heat by the glass cover. This was the lambda (Λ) shift theory to explain how heat could not escape from under the glass cover. It was the state-of-the-science then on heat theory.

The theory was soon discarded when sunlight was found to contain long (infra-red, IR) wavelengths naturally. Perhaps the knowledge that black colour absorbs heat and carbon is black advanced a new theory that carbon containing gases like methane (CH_4), and carbon dioxide (CO_2), so called greenhouse gases (GHG), contribute to a warming Earth. Without GHG, Earth would be cold. Too much GHG then would bring overheating as well. The theory treats GHG as a heat sponge in essence. Without any evidence as proof whatsoever, it has become de facto explanation for global warming to the extent that the 2008 Nobel Peace Prize was awarded to Al Gore. This chapter will **debunk CO2 theory** as wrong and scientifically baseless. We start by asking *"if CO2 and GHG absorb heat for warmth, how then would they release heat so Earth can cool down ?"* Is it not contradiction and wanting it both ways?

Black smokes billow up the sky in a fire. The blackness is un-burnt carbon particles but not CO2 gas which is colourless. Its rising up is from the heat of fire that expands the gas. On cooling CO2 sinks quickly to ground and disappears out of sight. CO2 is a heavy gas with a molecular weight of 44. Air has an averaged molecular weight of 28.9. Air is lighter than CO2. By Archimedes Principle CO2 would sink down to ground. Why do we not choke and gag if surrounded by CO2? The answer is that CO2 hides underground all its entire life. Its detection level in air is never above 0.03% (= 300 ppm). Check your lawn after a rain for signs of water puddles. Water seeps easily into soil until it finds bedrock or water basin that stops further sinking. Air and gas do even better. "Solid" ground to our _eyes_ is 100% porous to CO2. If CO2 is cutoff from sunlight by hiding in the earth, how can it cause warming ?

Water is the best molecular gymnast thanks to its unique electric dipole that can couple to **many** pulsating (E-M) radiation energy fields. Nothing else comes close to water in its heat capacity. By weight, specific heat of moisture is five times more than carbon dioxide CO2, which is barely better than air. Dry air is an insulator. Air in garments and blankets block heat loss. Air bubbles in snow keep igloos warm. Like a blanket atmospheric air shields a rock steady 15 C ground temperature from the hostile – 120 C cold of space. The enormous heat capacity of water makes it as the **only** true heat absorbing gas. Moisture in air not only absorbs heat, but also governs weather. Water is coolant in a circulatory system for **global cooling** as was shown in Chapter 10. Rather than called "greenhouse gas", moisture should appropriately be labeled *"ELEVATOR gas"* for its ability to raise and drop temperature plus its ability to rise up with solar heat and sink down by gravity pull. Its secret presence in Joseph Fourier's 1824 glass box has misled everyone ever since for almost two centuries.

Cutting to the chase, CO2 for global warming is utter twaddle. CO2 proponents have opted to overlook basic principles of Thermodynamics that at low temperatures the most efficient way to remove heat is a fluid circulatory system, NOT radiation. Such a system is found in cars, boats and even jet planes where cold air replaces water as coolant. Deep ocean current circulation is how Earth has tamed its own internal frothing inferno at 1050 C from reaching us on the surface for 4.6 billion years. If not for that, Earth life would have been long gone. Instead, ground temperature has stayed rock steady at 15 C unchanged. Radiation for heat removal is only a last resort when no other alternative exists and a high loss rate is necessary, but its effectiveness begins at 200 C at the minimum. <u>Earth never radiates nor could it</u>. CO2 debate is frivolous if radiation model for global warming is false. That said, it does not mean that global warming is fictitious. It is real enough, but not by CO2 or any GHG theory. The sooner we realize it, the quicker we can start working on real solutions.

The strongest condemnation against CO2 theory lies in its consistently low 0.03 % (or 300 ppm, parts per million) detection level in air. <u>Earth's real atmosphere extends to more than 35 Km way deep underground below our feet</u>. We all know where diamonds come from - elemental carbon compressed deep underground under extreme heat and pressure for million years. A gas like CO2 can sink that far down into isolated cracks and voids. To make diamonds, CO2 would be disassociated by the extreme heat and oxygen extracted would oxidize rocks and minerals, leaving carbon available for diamond making. Unlike molten minerals that flow and fuse into each other to create veins the way they are mined, diamonds emerge in isolated gems of varying sizes. This pattern agrees with gas pockets. 99.97 % of CO2 hide deep underground by the water table and bedrock where it dissolves into water to be absorbed by roots as building blocks for flowers and trees under osmotic pressure, to feed microorganisms like planktons in seas and oceans, to enhance coral growth worldwide like the Great Barrier Reef of Australia.

Greenhouse gas (GHG) anthropogenic global warming (AGW) theory is built on a rather flimsy simplistic or naïve <u>heat-in and heat-out</u> unsubstantiated assumption that a gas barrier can block heat from radiating out to space to cool Earth. After the heat-out part was proven bogus by arguments in previous chapters that explained how heat turn instantly into kinetic energy on a **fast** exit, the entire CO2 theory falls apart like a deck of cards. In respiration, it is oxygen-in/CO2-out for animals, and reverse in vegetation, but never the same gas both ways. Earth works the same way. CO2 theorists do not give Earth credit for her wisdom, resourcefulness and deserved versatility.

Solar Duvets (2)
In case a 500-mile (700 Km) thick atmosphere may be insufficient to insulate us from a murderous – 140 C cold space, Nature throws in two solar duvets of (2000 – 5000 C) hot plasmas as top and bottom flannel blankets as extra safeguard. They are the Ionosphere and Ozone band respectively to sandwich a dry insulating Mesosphere and Stratosphere

layers in between. Treat them like a double-wall uterus floating a fetus (Earth) in an embryonic fluid (atmosphere) in an ultimate protection*. Against these two superior solar heat firewalls, how effective would a low 0.03% (or 300 ppm) CO_2 "assumed" heat barrier be based on (GHG) greenhouse gas theory? CO_2 hypothesis is simply an illusion, fabrication, exaggeration, utter nonsense, outright fantasy but a scientific boldface wild conjecture at best. If this is not the last and final nail to GHG theory coffin, then what is?

*Note. As a protective mom, would she enjoy her embryonic fluid polluted, churned and her fetus subjected to high pitch jet engine whine instead of peace and tranquility? Is convulsion her likely reaction? Is intrusion any risk to her own health ? If she reacts, watch out !

Personally, I have no rankles with pioneer global warming alarmists Dr. David Suzuki or former US Vice President Al Gore. In fact, I admire and thank them for their courage and verve in raising public attention to AGW threat. They were only partly right, but also mostly wrong on its physics for explanation. Being not bona fide trained physicists that I know of, they cannot be held at fault. It is obvious that I am old-school in my attitude to science. The trouble with our modern media-centric society is that so many folks with idle time and megaphones in hand made available by technology offer unsolicited blue-sky opinions. Everyone seeks that proverbial 15-minute fame. TV and Internet have transformed societies from print readers to effortless visual viewers. The lens is ruling the world. Newscasts are embellished with infomercial noise and sound-bites to bolster their messages. That is our dilemma. Truth gets muffled under the overwhelming noise that would be absent in print media. Everybody is entitled to an opinion of course, but too few refrain from holding back their pride to be right. If winning is everything with big commercial benefit implications, it fosters tendency to outshout each other. Loudness cannot fortify falsehood any more than a grander edifice* of worship adds validation to a faith as far as the non-believer or believer is concerned. A more magnificent church, mosque or temple may glorify the initiator-builder for contributions, devotion, goodwill with psychological reward, but adds no value beyond visual impact. Overt hyping in self-promotion can be a serious distortion of truth. It is a line not to be crossed by ethical scientists. Ponds and wetlands need no advertisements to ducks and geese that would be naturally drawn to them like magnets for nurture and provisions. Give audience its credit for their own intellect. Competition for

R&D grants is one thing but once the data are publicized let the judging stand alone on their own merit. Our peers must be jurists and we accept their verdict unequivocally with dignity. Science is not a football to be kicked around in any political arena.

*As a side note and for the record, I neither subscribe nor support the controversial LHC (Large Hadron Collider) of CERN near Geneva from its outset. For their sake, I await to be proven wrong anxiously the sooner the better and wish them my best. I totally admire their dedication and envy the level of support they had obtained. At least on the surface it is a feat of engineering triumph so far.

Chapter 12

Global Warming Threat Assessment, Ice Shrinkage, and The 3-river systems.

Threat Assessment

Global warming conjures images of infernal Apocalypse and rampant desertification at its peak. Climate scientists have developed computer models based on that notion, which I consider somewhat far fetched and naïve. As long as there is water, that cannot happen. Earth is 72 % under water and the 6½ miles (or 10.9 Km) deep Mariana Trench near Japan is 23% deeper than the 8.9 Km Mt. Everest is high. I cannot think of a way how water can escape Earth gravity to vanish into space thereby depleting our stock.

Greenhouse gas (GHG) warming alarmists are mired in a HEAT model explained below;

HEAT model.
Heat received from the daily sun compounded by fuel consumption of human activity raises atmospheric temperature incessantly when heat release is blocked by a thickening GHG barrier like a cover. The truth is quite opposite when Earth has kept a rock steady 15 C ground temperature for as long as records are found. This heat model is clearly flawed. It should be discredited and replaced ASAP by the more credible STORM version below.

STORM model

This model embraces a mechanism of heat conversion to kinetic energy to explain how heat is dissipated quickly without raising temperature. This model uses moisture as heat carrier to regulate temperature and as energy conversion agent. In this model, total amount of Earth water is conserved and recycled by the mighty power of gravity. There is incontrovertible evidence for this model as the correct one as exemplified by our weather daily. Solar heat drives Earth climate through the complex action of moisture as told in Chapters 5 to 10.

Midwest states in southern USA have the highest incidents of tornadoes in the world by an incredibly 10,000%. In 2005, there were 1264 tornadoes compared to the next highest of 13 in Australia followed by only 3 each in China and Columbia. USA has earned deservedly nickname "tornado alley" of the world. Any day of the year there are 31,000 scheduled flights and between 5,000 to 8,000 airplanes in mid-air blanketing the sky. Its airport density of all sizes is by far the highest in the world. A 2007 statistics showed an aggregate of 15,000 compared to the next highest 4,200 (Brazil), 1,800 (Mexico), 1,300 (Canada), and 1,200 each (Argentina and Russia). Americans **fly** and drive far more often than others walk or bike each day. One has not far to go to see a tight correlations between tornado events and **flying frequency**. Worst of all is those hot jet exhausts up in the high atmosphere (notice vapour trails if you look up) that cannot be flushed down by falling rain because it is too high and too cold for any moisture to reach. Heat has nowhere to go but to build up enough to eventually whoosh down hard as kinetic energy in downbursts sweeping anything in its path like a toilet flush. Flying is a huge industry for the airline service and plane manufacturers. Therein lies a dilemma for the affluent west having to reconcile between global warming and the public lust for speed by air and by cars. Our biggest mistake is not having built more hi-speed trains like Europe and Asia. Each time a jet plane takes off requiring in the range of 1000 to 3000 gallons of aviation fuel, it burns up to **85%**[*] (proportionally less for longer trip) of its fuel for the trip to overcome gravity. Very wasteful. That is the root problem of an economy built on consumption, not conservation as it should be, and over-emphasis on speed. Haste makes waste. It is especially true for **short haul** commuter links between airport hubs and final small destinations **fighting gravity in numerous takeoffs**. It is as obscenely wasteful and ludicrous as heating up an entire pizza oven **each time** to roast a tiny hot dog. This foolish unconscionable practice needs serious curtailment immediately.

- Note. As a percentage of fuel actually used by trip. Airline safety regulation however requires all planes to carry **emergency surplus fuel** for either a return trip, or fly to alternate airports if weather prevents a landing, or congestion forces the plane on arrival into a waiting formation. The **extra weight** also costs more fuel consumed at take-off to oppose gravity. Flying is grossly fuel uneconomic. **The obsession in flying enslaves USA as an oil glutton and an outright abuser of fossil fuel resources. Flying ought to be restricted to inter-continental flights where there is just no easier alternative. From an energy efficiency perspective, land, sea and finally air transportation should be prioritized in descending manner to minimize fuel waste and ecological damage to the environment**

  Without this awakening in our collective conscience how could we morally deride Middle-East oil sheiks for treating their oil stockade as a private personal piggy bank when we just as immorally claim it as if it is our birthright to fly at will to the next town for a week-end visit, a quick round of golf or shopping?

Tall ship (SV Concordia, Feb. 17, 2010) flattened in 20 seconds in Atlantic 550 Km near Brazil, a deluge in Madeira Island, 2-feet snowfall in Moscow, blizzard in Dallas to Atlanta (a week before), Air France flight 447 plane falling out from the sky (June 1, 2009), 14 year old girl alone in ocean near Mauritius (June 30, 2009), adventurer-balloonist Steve Fossett (September 26, 2007), Egypt Air flight 990 (October 31, 1999), Swissair flight 111 off Peggy's Cove, Nova Scotia (September 2, 1998), and the many more we don't know because nobody was around to report are signs that Earth is gagging and choking to flush away her discomfort. Physical laws don't lie, compromise or even offer an opinion. When we act badly, Nature just reacts too (Newton's Law). Very simple and no fuss.

Chapter 5 pointed out the affinity between heat and water. For solar heat, different states of water have different coupling factors (impedance matching analogy) ranging in descending order;

1. moisture where separate molecules and their +/- dipoles resonate freely with many radiations to its fullest extent,
2. liquid water where molecules are electrically tied by surface tension, and heat of vaporization like an energy bar suppresses molecules from breaking free as moisture,
3. frozen ice with small air bubbles to reflect light, and heat of fusion as barrier
4. white snow with bigger air bubbles to be more reflecting, plus heat of fusion barrier.

Each descending one presents an increasing RF impedance to sunlight to determine the level of solar energy flow. Vanished glaciers and polar ice that have become moisture enhance solar intensity like a garment wearer switching from light to darker clothing to feel warmer.

Grain silos with poor ventilation are known for spontaneous explosions. Loose dusts mixed thoroughly with air ignite easily. It is the principle of a detonator to jar loose the explosive structure by a shock wave for maximal effects. From this perspective, human energy use is like a detonator and precursor to mounting global climatic violence. The process steps are;

1. Fuel burning raises air moisture level by subliming glaciers and ice caps for cooling,
2. Excess moisture raises solar intensity, albeit seemingly (by better coupling really)
3. Higher solar intensity boosts evaporation rate to raise moisture level even more and finally
4. begins a positive feedback loop.

A positive feedback loop in circuitry creates instability and eventual oscillations. Not nipped in the bud early, oscillation builds up amplitude from more energy until the power source is drained, the equivalent of developing a short circuit (Isc)*. Trouble is we have neither prior reference what solar source limit is, nor Earth's tolerance before reaching that point.

*Note. In photovoltaics (PV), each solar panel looking at sunlight has a maximum power it can deliver that is determined by its size and efficiency in a given sun. Without load, it shows an open voltage (Voc) that is maximum under that solar intensity. If output + and – leads ate shorted (unique for PV panels *only*, **DO <u>NOT</u> EVER DO THIS TO A**

BATTERY) the maximum current is (Isc) but with 0 for voltage that has been collapsed by the short.

Global warming threats materialize at two levels and in opposite seasons as follows;

1. Observable *summer horizontal winds* in the Troposphere like tornados, hails, hurricanes, torrential showers that cause landslides, and raging forest fires that form, grow and mature visibly by steps in hours to days with plenty of warning symptoms.

2. Abrupt *winter vertical AGBC ionic downbursts* high from the Ionosphere like Jet Stream and blizzards that can flush airplanes out of the sky such as Air France flight 447 on June 1, 2009 (approaching southern winter peak) flying from Brazil to Paris that fell into the Atlantic Ocean without warning nor a trace.

Ice Shrinkage

Ice is much more than the decorative item for your cocktail in the glass, or Kool-Aid and lemonade for the kids. Like an anchor that gives boaters reference and stability in an open sea, ice gives **both** temperature and density references to a global climatic system with stability and reliance. **Ice is our global climate guardian regulator and a thermostat**. All dynamic operating systems need references to run correctly. Install a defective thermostat in our car radiator cooler, the car would run erratically guaranteed. A fast spinning rotor, eg. windmill, that is without a governor will self-destruct at some point by its own internal stress. In global climatic terms polar ice and glaciers act as essential **temperature clamps** of a dynamic energy system.

Ice is a constant **_0 C_** temperature anywhere under ordinary circumstances and 92% the density of water. Anywhere ice is seen floating, rest assured that it is nine times bigger below that is obscured. Mariners stay well clear of icebergs for good reasons. Ice cover not only shields aquatic life below from hostile weather above, but also prevents water below it from freezing. The bottom interface between ice and water is always at 4 C where water density is maximum to make it sink down and away. Realistically, ice thickening is likely 99 % by new snow fallen on top that gravity packs hard into ice and 1 % by water freezing below. Ice drives a powerful downward convection current opposite in direction to a Bunsen flame heating a beaker of water, or sunlight that lifts moisture upwards to the sky. The mighty Humboldt Current emanating underneath Antarctica sweeps the Pacific Ocean bottom along a razor like edge of Chilean coast. It is the single largest liquid fluid circuit to nurture marine life and keeping Earth's 1050 C insanely hot internal magma cauldron in check like Genie in a corked bottle. Hilltop glaciers on the Himalayas, Alps, Rockies, Andes and Kilimanjaro feed our rivers for irrigation. Ice vitality to Earth life on land and in sea is indispensable. As long as there is glacial ice, risk of desertification as predicted prematurely by overzealous climate scientists will be low. Heat alone does not create deserts. Absence of replenishing water from glacial melt does without a doubt.

Vanishing polar ice and mountain glaciers are indisputable signs of global climate losing stability and heading towards chaos. They are valid reasons indeed to raise our attention and concern. Ice is a temperature buffer to combat global warming by absorbing heat in a warm wind for direct sublimation* to moisture. ***Evaporative*** cooling from water is too slow. A Bunsen flame to melt ice into water to raise ocean level is laboratory illustration only, which does not represent climate physics on Earth. Snow and ice are same material in hardness. Both are ice crystals. Snow contains more air than ice that is all. Blowing snow in gusty winds can sandblast glaciers away faster than melting ice by heat. One is mechanical erosion and the other thermal liquefaction. Heat in moist air converts instantly to kinetic energy on collision with a cold wind. The significant rise in ocean level as expected from ice melting never occurs, strongly suggestive that the sandblasting model (sublimation) is correct. Ice density is 92 % of water, its specific heat should therefore be less proportionally, but it is untrue. Ice specific heat is only 50% of water. It suggests that by lacking high surface tension like water, ice would be more apt to sublimate into warm air than melting into liquid.

*Note. Ice erosion model is consistent with how ocean waves have chipped away shorelines for eons since the beginning of time as disposal of all the solar heat ever harnessed.

Losing ice is like a car running on deflated tires without springs and shocks. Comfort and stability are compromised. A wilder temperature fluctuation (oscillation amplitude grows from instability and positive feedback loop) could follow. **Ice shrinkage is losing Earth's thermal guardian regulator or thermostat.** Open water does not offer equivalent buffer value for reasons below;

1. Surface tension retards evaporation until temperature builds high enough to break it.
2. Ice and snow are white by their trapped air bubbles, therefore reflective of sunlight..
3. Bulk water conducts and sinks away solar heat, retarding surface evaporation.

It would be instructive to **mentally** conduct four experiments (A, B, C, and D) below in order to better understand how different energy works differently. Two are heat based and two others rely on kinetic energy for erosion and evaporation.

A. Prepare two Petri-dishes side by side with each other. Put an ice cube in one and same amount of water by weight in the other. Warm them up on a hot plate and record their times of disappearance. Heat is conducted by the dishes. No doubt the ice will go last as it needs more heat called latent heat of melting.

B. Repeat the above but heat them in a microwave oven. Heat source is radiation that is absorbed faster by liquid hastening its evaporation. Heat radiation is reflected by the clear ice that slows down its loss.

C. Instead of heat, a gentle breeze at room temperature is blown over the samples by a fan. The reverse should occur with ice vanishing faster than liquid water. Lower vapor pressure of ice has advantage by accepting kinetic energy from the wind faster than the liquid water. Besides, liquid water molecules must overcome a high surface tension (73 dynes/cm) barrier before they are free. A direct change from ice to moisture without melting is "sublimation." Sublimation is an energy shortcut for ice . Recall also that ice is 8% lighter (less dense) than liquid water and floats on top.

D. Place a hot plate between samples and fan to combine heat and kinetic energy as warm wind on the samples before repeating the last experiment. Ice would go even much faster.

From these four mental experiments, we reach the conclusion below that a warming atmosphere from fossil fuel consumption;

mechanically erodes glaciers and polar ice by warm wind with combined heat and kinetic energy faster than surface evaporation from oceans and lakes.

As far back as 4 millennia ago, ancient Chinese and Greek mythologies acknowledged water as antithesis to fire. Water is the undisputed currency to control heat. In that currency ice is gold and money in the bank compared to water as copper and moisture as silver. Polar ice power drives our circulatory ocean currents to sustain a steady 15 C ground temperature while atmospheric moisture teetering between sun and gravity provides a livable 20 – 30 C temperature in air. There can never be too much ice for stability and assurance. Necessity of ice and glaciers in human survival cannot be overrated. We must do whatever is necessary to preserve ice. Terrestrial ice is the coldness of space transported to ground to underpin our victory over warming. **Ice to global warming is insulin to diabetes**. One without the other is demise guaranteed.

It matters little if one is religious or faithless. There is no clearer manifestation that water with its intricately intertwined properties through morphology from solid to vapor was meticulously engineered specifically to breed, nurture, support, protect and defend life in a mandate to allow our evolution. Missing any one link, life would have ceased long ago.

Currently our oil consumption level is 80 million barrels per day. That injects a lot of process heat into the atmosphere. Close to half of that is for transportation alone, especially wasteful flying in the affluent west.. In smoking the exhaled dirty air wafts everywhere as second hand smoke to pose health risk to non-smokers. Its infringement on the innocent public needs legal intervention. Fossil fuel waste heat is somewhat self-regulated with reduced impact to bystanders. Moisture is attracted to heat, which locks its location in a low pressure enclave. Moisture is lighter than dry air by the ratio of 18/29. Any time heat must exit fast, kinetic energy laden violent storms and destruction happen near its origin as observed in tornadoes and raging forest fires.

By all accounts, Copenhagen 2009 was an utter failure and a debacle. The lesson learnt was that AGW (anthropogenic global warming) based on carbon dioxide (CO_2) emission theory was on a scientifically shaky foundation from its start. An illogical premature CO_2 cap-and-trade policy arbitrarily pushed by economists or politicians offends our common sense as to how it could possibly offset fossil fuel consumption. It was nonsense and a mess was made. Copenhagen 2009 was example of a misguided pseudoscience finally facing truth. It typifies a tragedy when authority made and acted on wrong decisions without adequate debates nor consultations with the right professionals.

Using temperature to detect heat content is as futile as diagnosing cancer or diabetes by fever. Moisture as a wild card can masquerade heat content in air. The methodology used is scientifically amateurish and meaningless as using car speedometer to guess the engine speed (RPM) but not knowing transmission gear ratio. Nothing could be learned, and was not.

Over the past sixty five years since World War 2 ended in 1945 three significant events have occurred;

1. Global industrialization led by the technologically advanced west has given it a productivity advantage that created affluence and wealth accumulation.
2. World population doubled from 3 to 6 billions in direct benefit of industrialization
3. Migration of the skilled to the west to meet demand, or refugees fleeing political conflicts to offer themselves as low cost help

It so happens that the west (N. America and Europe) are both north of the Equator. We use fossil energy for food, shelter and transport everyday. Historically a 10% population rise creates a 30% rise in energy consumption. It is little wonder that the percentage rise in energy consumption has skewed towards the affluent west. These factors combined pose an extra waste heat burden on the North pole around which these activities cluster.

Winter of 2009 - 2010 saw Ottawa, Canada's capitol in a historic early spring turning the World heritage site Rideau Canal into puddles by January. Calgary was basking in warm Chinook for days when a blizzard was raging in Saskatchewan in between. What should climate scientists record as temperature ? Normally sunny hot LA was deluged in rain and landslides while would-be spring in Europe plunged into -35C deep freeze first time in five decades losing 200 lives. What temperature should represent such weather ? So temperature is wrong instrument to use in assessing heat energy *content* that is the issue.

There is a corollary between food and energy. In fact, food is energy. Anthropogenic global warming (AGW) is an energy ***content*** challenge like food obesity is to humans. Human activity is cause for AGW in the same manner as overeating is to obesity. I could not understand why climate scientists used temperature to detect a heat content problem when Earth possesses means to maintain its optimal 15 C ground temperature for years.

Our dilemma began with the misconception that Earth radiates away excess heat. It does NOT, certainly not at a lowly 15 C*. Its only means to expel waste heat is by turning it into kinetic energy embedded in air and wind to vanish on impact. Plenty of evidence is furnished by eroded shorelines and vanished polar ice and glaciers. They represent solar energy (like fossil fuel) intercepted by Earth since the beginning of time but disposed of.

*Note. If Earth radiates, humans would be daily recipients. NO way we could survive that scenario. Our body temperature at 36.7C is 2½ times as high as Earth's 15 C. In theory, we should radiate to Earth to cool ourselves as heat flows from high to low like water. It does not happen either way. Effective power radiation does not even begin until 200 C (473 K) minimum. Below that, it is too weak for detection without photomultiplier tubes to amplify its output.

The 3- river systems
There are literally 3 rivers flowing in Earth's ecosystem (***Figure 7***). They work in tandem with one another. River (1) is invisible for the most part has two separate and different halves, but is the source for the other two that follow. River (2) is obvious and familiar to us for swimming, boating, and all other aquatic activities. River (3) runs under the oceans. Any disruption of anyone of them would alter Earth's ecosystem dynamics to put life at risk.

River (1) is unconventional in that it has two separate actions. Its first half is an UP cycle relying on the sun and second half is DOWN cycle by gravity. The sequential push-pull action daily resemble our respiration of inhale and exhale, but of **heat** instead of air.

1A. First half (UP and Inhale) is solar evaporation mopping up atmospheric **heat** to lift it up the sky. As white clouds (micro-ice-crystals) heat becomes potential energy stored against gravity. Whiteness of clouds stops the solar process and blocks the sun.

1B. Second half (DOWN and Exhale) is clouds plus uncondensed moisture falling back to ground on cooling as either rain, snow, or morning dew depending on temperature. They are feedstock to drive river (2) directly or indirectly in a glacial melt-and-release manner. **As accumulated snow turns into in ice**, it powers river (3) from the N and S poles.

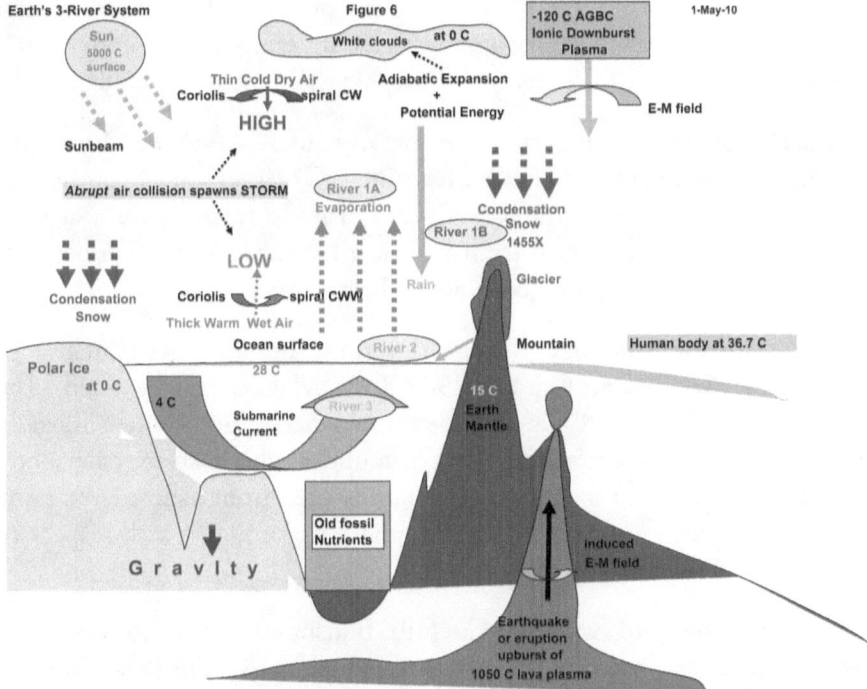

Figure 7. Excel graphic of "Earth's 3-River system"

River (2) is our familiar irrigation water flowing in channels and canals for farms and marine navigation. Without its irrigation, sun baked dry deserts would spread. Clearly river (2) is a keystone to life. If it stops, we starve and perish.

River (3) emanating underneath polar ice circulates our mighty oceans to keep Earth cool like a car engine system to maintain optimal temperature for best performance and providing comfortable interior climate to occupants. Only permanent polar glaciers can sustain river (3) flow. Loss of polar ice is like your car engine minus its water pump.

Topographically (2-D view) Earth is 72% oceans and lakes. Interface between land and sea in fact is 3-D. Underwater gorges like Mariana Trench near Japan can be deeper than Mt. Everest is high. This raises the interface ratio closer to 85 or 90 %. It means that ocean currents dominate Earth's cooling of its internal smoldering 1050 C lava core (the car engine) and keep its surface at a livable rock steady 15 C. Earthlings are sandwiched between an enormous microwave oven (sun) from above and a BBQ flame (lava core) below. Without an ocean circulatory cooling system, we would be done. Water obeys law of gravity flowing from high to low. Heat does the same. Atoms at high temperature with more kinetic energy than those at lower temperature pass their energy by collisions.

Human body operates at a critical temperature of *37 C(=98 F)* with a tight tolerance of +/- 2 C each side. It is a fever at 40 C (=104 F) or a hypothermia at 35 C (= 95 F). Either one is fatal. It is easier to retain body heat with clothing than repelling heat from hot air all around. River (1) does that by cooling air to 20 - 30 C (68 – 86 F), below our body.

Global warming is no joke. It is a very serious. Ignorance of climate jogs my memory back to an October 1987 Time Magazine story titled "The deadly Brazilian glitter" when a 6-year old native girl rubbed illegally discarded radioactive material she had found in the bush over her body to make her glow like a goddess. Her family died from cancer later, bone marrows destroyed. Aren't we doing something similarly with respect to our climate ? Ignorance has no bounds.

Do not take the word "warming" literally. It just means rising heat content. Temperature can be manipulated by evaporation as every nurse and athlete knows using a wet cloth. It is why the UN-IPCC body was unable to wrestle the problem down by temperature data alone. Its approach was scientifically flawed from the start. Heat is atomic activity. Animals need higher temperatures for mobility. Stagnant and immobile, vegetation stays cooler as long as its internal water does not freeze. Global warming will NOT bring infernal Apocalypse as ignorant propaganda alludes to, thanks to the copious supply of water on Earth. Instead, severe climatic violence in air, at sea and on land will manifest itself in higher fatality plus property damage when heat exits fast as kinetic energy embedded in any media from which it is expelled. Air France flight # 447 from Brazil to Paris in June 2009 was flushed out of the sky by an AGBC ionic downburst originated from the Ozone band most likely. 2009 - 10 being El Nino year, an extra wide Jet Stream pushed the entire winter system farther south depositing our usual Canadian snow in the USA and Western Europe. A surprise deluge in February 2010 in the Madeira Islands in mid-Atlantic Ocean came out of nowhere. That is why some environmentalists prefer the moniker "climate change" to avoid any temperature implication.

At the time of this writing, there was a consecutive series of earthquakes in magnitude 7.8 on Richter Scale in Haiti (January, 2010), 8.8 in Chile out in the ocean (March, 2010), 6.4 in Taiwan only 3 days later, 5.6 in Cuba (March 20) ,7.2 in Mexicali east of San Diego in California (April 4) and most recently (April 6) 7.8 in Aceh province in North Sumatra of Malaysia. Their close proximity suggests high reverberations in Ionosphere. Are they early indications of instability oscillations ? Blizzards continue to wreak havoc in places that have not seen snow in decades. To me, these are unhappy signals sent by Ionosphere as precursors of what more might likely follow. It would be foolhardy to shrug them off as random events. They deserve our close scrutiny in order to learn more about them.

Radiation physics obeys rules that apply to all high frequency RF circuitry. Moisture in the sun-earth radiation circuit represents a capacitive reactance ($1/\omega C$). In RF circuitry high capacitance drops reactance lowering circuit impedance to enhance a higher current flow (more absorption). Darker colour absorbs solar heat faster than white while sunlight stays constant. This is the real risk in global warming. If moisture continues to rise (from ice depletion) solar heat would be absorbed faster triggering more moisture sublimated from ice as Earth strives to regain temperature control. This sets up a dangerous **positive** feedback loop that brings eventual instability and violent oscillations in an attempt to seek equilibrium. More severe climatic violence might include horrific traffic accidents in all forms including sudden AGBC ionic downbursts in aviation, unexpected ocean rogue waves and squalls that cut road visibility. The possibility is endless from bringing down power grids, rupturing pipelines, destroying infrastructures, blinding blizzards, abrasive desert storms or raging forest fires. If severe shoreline erosions should occur, it might tip the delicate balance between abutting tectonic plates in theory at least. In energy use, the amassed heat has to exit and it is wise to release it slowly under control, where possible, than in a fast violent outburst. At its worst should oscillation kick in, Earth can no longer be our home. Let us wake up before we reach that final threshold and it will be too late then.

On April 20, 2010 a methane fire broke out on an oil rig Deepwater Horizon operated by BP (British Petroleum) 40 miles offshore from Louisiana coastline in the Gulf of Mexico. It burnt out of control for two days before sinking 5000 feet under the sea with three broken pipes spewing thick crude oil continuously that produced oil slicks on the surface. A week later, the initial leak estimate of 1000 barrels a day was revised up to 5000 barrels* (or 210,000 gallons/day). Winds shifting from off-shore to onshore threaten sensitive shrimp, oyster, fish farms and a natural habitat for millions of migratory birds in the pristine Mississippi Delta. Attempts to plug the deep sea oil leak, and boom confinement on the surface in a control burn-off have failed. Prospects of an unstoppable environmental catastrophe plus its dire economic implications loom larger each additional day that experts say might last for months before a nearby relief well can be drilled. US Coast Guard is on record saying the incident will make the 11 million gallons oil spill from the bowel of tanker Exxon Valdez run aground in 1989 in William Sound look like a mild cough or stomachache. I see this as bad omen of what to come.

*Note. Experts viewing a BP provided underwater video feed 2 weeks after the incident have challenged this estimate as likely close to 70,000 barrels per day. It is a discrepancy of 14,000 % !

Over the eons, ancient civilizations have mysteriously vanished without written explanations. In business there is the Peter's Principle whereby an individual keeps promoting him/herself to beyond his/her natural level of competence and begins to flounder. Perhaps civilizations work the same way towards self-destruct by over-reaching and over-assurance or confidence in its technology. Our addiction to plus an insatiable appetite in fossil energy, and reluctance to moderate our ways will lead to that inevitable outcome by hitting the proverbial brick wall that stops us because we refuse to stop before it. Our best legacy to future generations is not in WHAT material or wealth to leave behind, but in HOW to live an exemplary sustainable wonderful life in order for them to see and enjoy their offspring like we do ours. Pass them our valuable learned experience, not the morass that we have created in our greed for profit, speed, comfort, disregard to others, selfishness and utter foolishness.

Chapter 13

Clean energy, Solar, Wind, Bio-fuel, Remedy & Solution, Inferences and Concurrency

Clean Energy

Clean or green energy is a misnomer and a euphemism. Energy is heat that is necessary for activity. **Process heat itself is waste**. Radiations can propagate as pulsating (E-M) energy fields, but heat needs material for transport. Spent heat must exit a system to let in fresh energy to perform new tasks. It cannot be helped. Greenness may mitigate its negative impact a bit but it is not benign. Greenness is a self-soothing illusion to assuage a sense of fear, to delay the inevitable in faint hope of a better solution in future by more advanced technology. Sprinkle on top some political gamesmanship, marketing savvy and consumer gullibility, we have a perfect witches brew for colossal disaster in the making. Rather than waking up to the realization of being down the wrong track altogether and instead of turning around while there may be time still, greenness propels us further by beating the bush trying desperately to find a short cut through even less known terrain. The intention may be as noble as introducing a new life saving drug, but the assumed risk is equally high when side effects are yet unknown and gathered data are too scarce to base on for a decision. We may hurry into really bad surprises.

Energy can be either as-produced, like solar wind and geothermal, or from storage. Line of demarcation in between is blurred. Hydro, wind and tide look as-produced but are in fact solar storage converted to kinetic energy same as fossil fuel that in addition has been long compressed by heat and pressure. Renewable energy is recurrent on a schedule and tap-able only then, therefore limiting its availability and by extension consumption. Renewable alternative power slows depletion of our fossil fuel stockade in a trade-off for slower waste output, but in effect settled for less productivity. In any given process, output has a tight correlation to energy used. If not, we get into quandary of something for nothing.

Except for geothermal heat from Earth's inner core, all surface energy harnessed by Earth comes from the sun. The sun heats up land, evaporates oceans and lakes to power wind and waves by kinetic energy. Photosynthesis and sunlight grow plants into fossil fuel storage eventually. The energy cycle begins with solar radiation and ends as kinetic energy in wind (air) or waves (water). Heat is **activity** energy in between them, and <u>water its transporter plus conversion agent</u>.

In the energy family hierarchy, electricity is big brother while kinetic energy is kid brother to heat. As shown in earlier texts, heat converts to kinetic energy readily by a fast freeze of warm moisture, creating a storm. Seen this way, fast freeze does the same as fuse to a bomb. Once lit, it is instant explosion (kinetic energy). Heat to electricity conversion is uphill harder, and not very efficient yet. More commonly done is kinetic energy converted to electricity by a rotary turbine. Wire coils spinning (kinetic energy) in a magnetic field synthesizes, as it were, pulsating (E-M) fields to move as electricity (not electron flow, see Chapter 14 next). Only low frequencies are achieved this way and electricity needs material for transport just like heat. Radiations are exclusive to very high frequencies that are impossible to produce by mechanical means because of momentum considerations and highly dangerous destructive centrifugal forces.

Solar

Application of this energy taps into the beginning of the energy cycle as it enters Earth. Solar energy is **extracted** on demand from a pulsating (E-M) energy field that originates from 93 million miles away by resonant coupling with **water** on Earth. A waterless Earth, like Moon, would be stone cold despite 24/7 sunlight. That is true of Antarctica that is bone dry (moisture frozen). Solar energy obeys laws of radiation physics. There is no deflection or redirect with radiation if resonance is absent. Only particles do. Listening to radio news or watching a TV channel, signal from the station is coupled to your tuned receiver by resonance. There is no stream of energy particles shooting from the station to your receiver. The energy is **drawn**, not delivered by sender. In a sun baked desert, solar energy is reflected by the non-coupling sand. The heat felt is from tiny residual air moisture in resonant coupling being crammed with a lot of heat (high heat density means high temperature). A really dry desert injects ZERO solar heat into the environment while a wet swamp of equal size at same location would upload a large quantity of solar heat but not felt as hot, muggy perhaps.

Covering farmland with dark solar panels to generate electricity, we have altered its energy dynamics. Solar energy couples faster to dark (solar panel) surfaces. Photovoltaics (PV) was developed in the 1950's by Bell Laboratories for satellite power to compete with the former Soviet Union, who used onboard nuclear reactor, in the hot space race at the time. Terrestrial PV is ideal and cost effective for remote standalone power stations especially in telecom repeaters. PV grid connected utility power to augment supply is increasingly implemented in the quest for reducing foreign oil import but is controversial in terms of land use, and questionable in cost effectiveness. Standard production grade silicon solar cells are at best 15% efficient. At grid level it is more like 11 %. There is a tradeoff of 89% extra waste heat at the solar farm that would not otherwise be in its original state. Did greenness anticipate that? PV infrastructure cost is 90% upfront with around 15% annual maintenance cost in replaceable parts, mostly batteries or electronics for life warranty of 20 years against 10% output degradation. If the 89 % waste heat is not extracted for secondary usage like hot water for laundry, car wash, preheating baths in hotels, and similar applications to speed up investment recovery, then PV grid power is inflicting unintentional environmental harm to production site by moisture extraction out of the ground rendering it arid dry and prone to soil erosion in time.

Wind

Wind energy is solar energy salvage before final exit. Un-captured, it dies on impact. Wind is a renewable alternative but its site choice requires thought, deliberation, sacrifice and accommodation from neighbourhoods that are affected. Windmills emit audible drones and sub-audio (below sound frequencies) powerful vibrations to rattle objects that contain high water content. As liquid, water molecules lattice together hexagonally (as seen in snow) into heavy molecules that resonate with sub-audio frequencies used in submarine communications by the navy. Human being is over 80% water and sensitive to sub-audio drones. A minimum distance of several

113

miles should separate animals from windmills. There is the extra risk of rotary blade failure and inference with seasonal bird migration. How is greenness doing here ?

Bio-fuel

This notion adopts the same psychology of diet beverages, sugarless gum, low fat, low salt fast food and other pretend compromises that are more to opiate the conscience than to satiate the palate in a devious dishonest self-delusion in the conduct of a misdeed. It has the inherent danger of fostering more recklessness under false pretense of a damage-free impunity. In my opinion, it is too high price to pay for an addiction. It is wasteful of precious resources. Carbon is the essential ingredient in combustion to produce heat. Low carbon is low heat output. To accomplish the same task using conventional oil, extra low carbon bio-fuels need be burned that might obliterate any potential cost advantages without reducing heat waste output at all. Land, labour plus solar energy dedicated to its growth should be economically better used for food production. If farmers are enticed by a higher commodity price to grow fuel to burn than food to eat, then we have a serious societal and ethical problem. Extend the argument to doctors and nurses, who cost us a fortune to train and educate, opting to enter into cosmetic surgery and implant clinics for better income rather than saving lives in hospitals. We need to maintain a proper priority and set of values as a community if we were to continue. Cars, boats and jet planes require high energy fuels like diesel and gasoline that were compressed for million years. How can they be artificially duplicated ? Besides, if CO_2 is refuted as global warming agent and carbon footprint is no longer relevant, then CO_2 emission is in effect no longer any issue, thereby rendering bio-fuel debate moot altogether.

CO_2 Sequestration

This would be purely a redundant made-work scheme like hiring folks to return earthworms to soil after a rain. It is even more unnecessary and pointless now with CO_2 theory proven incorrect and irrelevant to AGW.

Remedy and solution

Energy is like the mythical Genie in a bottle. His magical power gives us the illusion as our slave. Yet he could at any time turn on us and we are unable to control or coax him back to the bottle to cork it. Our dependency on him leaves us vulnerable. Do we desire to be a hapless zoo animal trainer who at any time could be victim of his/her beast in a flash of anger or impulse ? How worthy is that risk ?

There is a corollary between energy and glucose. In fact, glucose is energy. Overeating causes obesity that leads to hypertension and diabetes. It behooves us to be astute, vigilant and knowledgeable about food intake in order to keep obesity in check. Same attitude should prevail in energy use. Short-term satisfaction that inflicts long-term harm to Earth's environment is foolhardy. There is no free energy ride. Like a fetus, our energy load is lumped on Mother Nature's own burden to shed. She would do whatever necessary to clean house, including violent weather, earthquakes and volcanic eruptions to maintain her equilibrium and sanctity. There is a hefty price for us to pay should we not take care of her first. In a sick mom, fetus (we) in her womb suffers with her.

Long distance jets like to fly in the rarified weather free Troposphere-Stratosphere zone to save fuel and ride the Jet Streams when available. Over continental USA 31,000 lights are scheduled daily with 5,000 to 8,000 planes blanketing the sky at any instant. It is an obscene level of intrusion to a pristine upper atmosphere that acts like our life blanket.

To slow down global warming requires a 2-tier solution in recognition of the differences of their origins, energy intensity and seasonal variations. Perhaps up to 30% slowdown is achievable by implementing ASAP (as soon as possible) the following measures;

1. ban short haul flights under 1000 miles and replace them by land transport that do not fight gravity.

2. mandate domestic flights to a 10 Km ceiling to terminate heat deposit above Troposphere. Fuel saving from higher altitude flight does not justify extra climb against gravity.

3. mandate flight cruise speed reduction by 20% at least. Wind resistance obeys a velocity cubed (V*3) law. Doubling speed burns 8 (=2 X 2 X 2) times more fuel. A 20% speed reduction burns *only 51.2%* (=0.8 X 0.8 X 0.8) of fuel at full speed before, saving *half* the fuel but arriving just 20% later. A 4-hour flight becomes 5. Time in exchange for saving fuel and environment. We need a new paradigm.

4. Likewise drop highway limit from 55 MPH now to 50 MPH maximum, just 10% slower yet for an additional **25%** fuel saving. It takes 11 minutes longer to cover every 100 miles of road. It is time to decelerate, now with more cars on the roads including green hybrid cars of lower speeds.

5. Shift production activities from winter to summer where possible. Energy for winter survival is inevitable. Anything else would raise air moisture that raises solar absorption in the Ionosphere. Dry cold air at 0 C is 9% denser than dry warm air of 25 C. At equal speed, winter wind deals a bigger wallop and more devastation. In truth, the difference would be even higher as warm wind sucks moisture that makes it even lighter again. We should aim for least winter storms that erode polar ice and glaciers that are thermostats.

6. If winter production must proceed, it should be done in tropical or temperate climate zones to facilitate its waste heat exit assisted by the sun.

7. An inter-governmental plan to relocate senior citizens and retirees in winter to warm regions with health care coverage. This saves not just heating energy but also snow removal where they would live to effect a total area shut down.

On April 15, 2010 a volcano under a 200 meter thick Eyjafjallajokull Glacier in Iceland near Skogar erupted. A normal volcanic eruption like those in Hawaii spews hot molten lava in spurts that roll down mountain sides. Punching through a glacier however, lava heat instantly turns ice into steam 1500 times bigger in volume in a gigantic explosion. The net effect is a sudden pulverization and condensation of molten lava into ultrafine shards of opaque glass and obsidian dusts blanketing the sky kilometers high like thunder clouds that poses a major hazard to aviation. A night flying British Airway Flight # 009 four-engine 747 Boeing flying through a volcanic eruption near Java in Indonesia in June 1982 lost all engines and glided powerless for 14 minutes before restarting them luckily at a lower altitude to land safely in Jakarta. The Icelandic volcano eruption closed Northern European airports for more than a week preventing an estimated 600,000 passengers from travelling anywhere. Direct cost to airline industry alone is $200 million per day. Countries began planning cruise ship charters and naval vessels to bring back their stranded nationals.

It is as if Mother Nature was fed up of rude intrusions into her pristine high atmosphere inner sanctum and sent out warming that "enough is enough". She is upset.

On August 27, 1883 when Equatorial volcano Krakatoa in Sunda Strait blew itself to smithereens before sinking below the ocean, its explosion was loudly heard over 7,000 miles (or 11,000 Km) across Indian Ocean (sound in water travels 4¼ times its speed in air of 741 MPH) from Brisbane in Australia to Zanzibar, East Africa, sent waves around Africa into the English Channel over 11,000 miles (or 20,000 Km) away. The blast produced a 150-ft. tsunami devouring everything in its path. Experts calculated that its energy sent pumice 25 miles (or 40 Km) high into the Mesosphere. Without wind to disperse, the detritus lingered for years at the mercy of gravity as the only mechanism for it to return to ground. In Troposphere below, wind spread fine dust all over the globe to create a nuclear winter to last more than two years. Europeans reported golden sunsets every night as temperature sank 4 C below normal. Intrusions into upper atmosphere leave long lingering memories as lasting reminders of its disapproval, revulsion and rejection..

Ultimately, we must embrace without reservation the Buddhists' maxim of **Restraint, Moderation and Propriety** as golden rules in energy consumption if we are sincere in leaving behind to our beloved children and theirs an enriched and safer world than the one we have inherited. Our best legacy to them is less value in what material to leave, but more in modeling a proper lifestyle to follow from our experience.

Environmentalism is neither a fad nor business promotion. It is plain common sense and basic animal instinct. It is hard pressed to find an animal that would defecate or urinate its own den that is shelter for nursing their young. We have two pet cats in our family. Over a decade since they have been with us, not once did they stray from a litter bin that is theirs for bathroom functions. They just visit without coaxing. It is that natural. Human beings have to be deranged at worst and insouciant at best to cast wanton wastes in their own backyard indiscriminately and ignore consequences in search of a fast profit. We foul it, we live it. We must not do secretly in public what we would not do in our private home. Expand the notion of neighbourhood. It is all about attitude and vision. However long our civilization may survive and whatever technology we may have in our control, we are forever the fetus in Mother Nature's womb suspended in her embryonic fluid. A sick pregnancy means a baby at risk. It behooves us to protect the nursing mother in order to save ourselves. The abundance around us enriches us. Without it, we are lonely poor orphans living in misery. A shared world is so much more fun except for the apathetic, emotionally disconnected, and devoid of social conscience or naturally disagreeable who cannot be pleased no matter how. Protect others in order to preserve us.

I am not a social scientist, but my training taught me logic. Logic helps develop insight to predict future a lot better than wild guesses. Frankly, our addiction to fossil fuel has crossed over to **abuse** long ago. Abuse is beyond caring. A sense of entitlement has taken hold of the incorrigibles, especially in flying just to be there fast*.

*Note. After the 9/11/2001 NY World Trade Tower collapse, flying is no longer a pleasant nor a speedy experience. Extra airport security measures and multiple delays have turned flying into a tedious bore for frequent flyers. With volcanic ash thrown on top, airlines are losing money by the barrels. Audio-video real time teleconferencing has gained favour in its place to save time, money and **energy** for the business community. Thank goodness for that. Hallelujah, I would say ! Tourism has suffered as well.

Besides a nasty worrisome Gulf of Mexico BP rig oil spill, an equally ominous front page news in the opening week of May 2010 was daily riot in Greece over austerity measures imposed by government in exchange for a $140 Billions economic bailout. The political and economic instability threatens the entire EU community. Energy abuse is not an isolated issue. It is part and parcel of a spoiled dysfunctional run-away so called Boom-Echo generation, and for which the Boomer generation (yes, with me in it) is mostly responsible. Boomer generation is more prolific in inventions, discoveries, creativity, productivity and wealth accumulation than any other generations in recorded history. In the rush for profit, we have over-provided our next generation and in so doing depriving, not helping in fact, them to be equally creative for themselves. Boomers inadvertently have desecrated under the illusion of progress institutions that were built by hard work of predecessors. Not having to walk blocks to find a payphone and put real coins in it for a call home, kids today abuse that invention with texting or sexting as their birth right. Without money for toys, kids would be forced to learn to develop their imagination for games and fun. Store bought video games might make money for the professional game designers and sellers, but stifle the imagination of the budding young consumers. Likewise fast food outlets everywhere obviate the urge to learn to cook when hungry. Even a most fundamental drive to survive has been compromised. Supermarkets distort our children's notion from where produce come, apparently in the cold storage at the rear instead of farms miles away. Some elementary school kids when shown tomatoes cannot name them. Over-fertilization is the principle behind weed-killing by making the weed grow **too fast.** that it wilts. Over-coddling our children cripples them in self-esteem and motivation. Over-indulging them is killing them like weeds. Not only have I concerns of their ability to fend for themselves, never mind looking after us in our autumn years, I am troubled by the dilemma confronting them while battling withdrawal symptoms when reality lowers the real boom on them unprepared one day. How will they cope falling down hard? They will need rehabilitation like quitting smoking or drinking, or intervention for drug abuse. Energy abuse is of course a scourge of the affluent that does not affect the poor, who will continue to eke out a meager living if they can like they have always been. Their chances of survival look a lot more promising under that scenario than the rich and spoiled. Ants and critters made it but all those mighty and powerful dinosaurs at the top of the food chain perished in the last cataclysmic episode.

Inferences

Based on arguments and suppositions presented thus far, pending verification naturally, it is reasonable to infer the following for a new level of investigations.

1. abrupt AGBC ionic downbursts originate from either Ionosphere or Ozone band that are plasmas
2. both are hot ions in nature but having converted to high speed cold downbursts and Jet Streams.
3. due to greater height and thickness, Ionospheric downbursts are far more powerful than Ozonic ones
4. Ionospheric downbursts experience more (B X V) deflection to spiral to the N and S poles (ccw and cw respectively)
5. Ionospheric downbursts become Jets Streams in upper atmosphere as polar and sub-polar winds.
6. Ozonic downbursts subjected to less (B X V) magnetic deflection plunge vertically to spawn tornadoes, etc.

7. molten magma as liquid plasma reacts to Ionospheric ionic magnetic flux to induce Earth's magnetic field
8. Ionospheric or Ozonic ionic downbursts at a tectonic junction or fault line could trigger a quake or eruption
9. atmospheric ionic downburst induces a liquid plasma up-burst from lave core in a counter magnetic response
10. A more agitated Ionosphere and Ozone band is potential for higher and more often earthquakes or eruptions
11. Earth's inner energy reserve has lasted so long because the energy origin is external, namely solar
12. If a spring is compressed to bounce, the energy is from the compressor, not from the spring that just returns energy

Concurrency

For lay climatologists and geologists like me, there is a possibility to either verify or disprove AGBC ionic downburst proposition on paper first without elaborate laboratory setup. Under AGBC theory surface storms, eruptions and quakes are triggered and energized by AGBC ionic downbursts that share a solar origin as common trait. If it were true, one particular AGBC ionic downburst can trigger **only one event alone** to dispense 100% of its energy, but not two. Call it the "Concurrency" test. Simply put, a surface storm cannot happen with an earthquake or eruption simultaneously in minutes, inside an hour at best, at the same location.

In theory, if quakes and eruptions are provoked by lava core activity inside Earth independent of outside AGBC trigger, then concurrency with a surface storm is mathematically possible by pure chance. The long life (days) of hurricanes compared to abrupt quakes and eruptions makes them lousy time markers for concurrency tests, but sudden and equally abrupt tornadoes are a perfect match as time markers.

If we scour old records of tornadoes, quakes and eruptions by date, time and locations, all we need is only ONE perfect concurrence to **disprove** AGBC ionic downburst theory. It does not prove it either, but will deal a blow to its possibility in a backhanded way. The Caribbean chain of 700 islands and Indonesian Archipelago of 16,000 islands are renown for both earthquakes and hurricanes (less for tornado unfortunately). They would be ideal candidates for this test.

We hear of storm season due to the Earth location in its annual orbit, but is there earthquake or volcano eruption season ? If not, why not ? Are they truly random ? I overheard that 130 quakes happen yearly as a statistic. If a quake or eruption season exists, then there is a tie to the sun and its influence to bolster the AGBC theory somewhat.

As arguably intelligent human beings, we simply have two diametrically opposite choices to make as follows;

1. continue with self-abuse in food, energy and consumptions in defiance with a sense of entitlement and let Nature take its course for whatever consequences may bring in dream of a future technology that solves all our immediate woes, or

2. judiciously take charge by moderating our ways to harmonize better with Nature to leave next generations a sustainable legacy with conservation, vigilance and propriety by curbing waste and sharing more with others.

Abuse in one form or another is so rampant today that homeless alcoholics have taken to quaffing mouthwash like Listerine to get whatever little alcohol there is. Robberies in which easy open cash is overlooked in favour of cigarettes or analgesics like OxyContin are a new but persistent phenomenon. Black market contraband smokes and cannabis are widespread. How soon will it be before black market gasoline will join their rank at our present voracious consumption? Criminal elements no doubt have it in their cross-hair waiting in glee and anticipation. They specialize in that sort of "business development". OPEC has lost its monopoly long ago to renegade oil wildcats. Politician are at wits end deciding to drill or not drill further out in deep seas to keep a price lid on dwindling home based oil supply in desperation to appease an insatiable nation, all the time knowing full well and waiting for the last straw to fall and break the camel's back. How did we come to this? Nobody before had the gumption nor courage to lead and put the brakes on to lose popularity. Why do leaders not lead when they are elected to do so ?

It is entirely up to us. Use the Genie taking the risks or keep him bottled and enjoy peaceful sleeps without worry.

Conclusion

Without AGBC ionic downbursts, there cannot be earthquake, volcanic eruption, Jet Stream, tropical storm, blizzard, glacier, polar ice cap, airplanes falling out of the sky mysteriously nor even freezing winter. Earth would be Utopia seemingly. That said, but it is impossible and inevitable, like death. Besides, without polar ice caps and glaciers as Earth's thermostat and frozen water storage, there would be no submarine ocean currents to feed the bottom of our food chain. Life may be even worse then. On balance, we might have got the better bargain overall in fact.

Chapter 14

My Definition of Energy, Electricity Flow, Quantum Mechanics and Relativity.

My Definition of Energy

Dictionary defines energy as "ability" to perform work. Ability is a descriptive term but not a scientific explanation. Students need an "analytical" definition like "water consists of two hydrogen atoms combined with one oxygen atom" to get a clear picture.

Science has given us a definition for

Work (W) = Force (F) X Distance (D).   (1)

The nature of distance D is open. It can be straight (linear) or curved (non-linear) but not cyclical (rotary). For rotations, the definition is modified to;

Work (W) = Torque X Revolution   (2)

If we try the Newtonian approach of treating tiny increments of distance (D), we might reach a scientific definition for energy. Work is the sum total (integral, $\int$) of energy used in a task by a force (F), over distance (D) during the process. Mathematically it looks like

Work (W) = $\int\Delta$(W) = $\int$F.dD = $\boldsymbol{\int m(dV/dt).(V.dt)}$ = $m\int$V.dV = $\frac{1}{2}$m.V*2
= Energy.   (3)

It is instructive to look at the middle term (italicized) for an interpretation.

The term (V.dt) suggests a guided motion in the event velocity (V) is function of time (t). From that, we can define that;

Energy (E) = Force (F) X Motion (M).

In the energy definition, the word "motion" replaces "distance" in the work definition. Work is a sum total after it is done regardless of HOW. Time taken and whether done in discrete steps or in one smooth flow are irrelevant. Motion implies a seamless ***action*** from start to finish in a time continuum to preserve the energy character. Motion is a key component of energy. Energy denotes action, while work is just an end result.

This energy definition applies to matter where motion is visible. With non-matter like radiation, energy is a pulsating high frequency electromagnetic (E-M) field. Neither energy nor its motion is visible, the definition is modified to;

Radiation (R) = E-M Force *Field* X Transmission (to a target)

Both definitions demand **Force and Motion simultaneously.** One without the other cannot be energy. Energy implies action. Action requires pressure to create motion either of material or something intangible like radiation. Following pressure created, it is still NOT energy until accompanied by movement. In circuitry, voltage (V) is the pressure only but current (I) that flows (motion) is the true energy. Radiation is E-M force field in motion.

Enterprising folks with perpetual machine fantasy invariably fail the motion test as soon as a load is put ON. Their inventions (rotation in all cases) stop upon loading. They are fooled by a magnetic field that gives the force, but not motion. Energy is in the motion.

This definition validates my argument that the Sun will last far longer than predictions made based on its computed burn-rate. Its powerful E-M energy field might reach far corners of the solar system and beyond, but its radiant energy only flows to objects that can receive by resonant coupling. There is no energy lost to empty space like streams of particles (photons) would. Dreams of sails for space vehicle propulsion to capture solar photons are pure fiction, poor imagination and utter scientific nonsense. Photons are a mental construct with zero mass, and carry no momentum for motion.

Bad science like CO2 global warming theory that has fooled us for 186 years since 1824 needs be stopped.

Electricity Flow
In circuitry, current flow was assumed to be free electrons physically zipping down the wire like particles to deliver energy. I think this visualization is wrong. Some suggest that electrons do not zip from end to end at the speed of light ($c = 3 \times 10^{*}8$ meters/ sec., or 7.5 times around the Equator) but instead bump each other by impact down the wire to move energy. Trouble is that no matter how lightweight an electron is, it still needs a finite time of acceleration to achieve speed of light. So movement of electrons is not it.

The only thing that reaches speed of light (c) **instantly** is an energy field that has no mass. Think of the wire as a long coaxial capacitor of high capacitor value (C) with all mobile electrons as one negative electrode spread on the skin*, and all the space charge (fixed immobile positive charged atoms) as one positive electrode in the wire core separated by a gap in the order of one atomic radius (a few Angstroms). A large capacitor would present a low impedance ($1/\omega C$) to low frequency currents but a short to high frequencies. This is why simple wire conducts well at low frequencies but presents a problem to high frequencies, unless it is coiled to include some inductance (L) for voltage. This visualization reinforces the argument that current flow is in reality a pulsating (E-M) force field transmitting down the wire at the speed of light like a waveguide. The inference from this is that all electricity flows by (E-M) force fields, not electrons.

*Note. "Skin effect" is a well known in high frequency RF circuits where current flow is proven to proceed along the outer skin without going through the conductor inner core.

Quantum Mechanics and Relativity
Technically, neither topic belongs in this book. Upon nearing completion however, I decided to make brief reference to them in the context of perhaps a last parting comment to those yet unborn grandchildren for them to get a glimpse and basic understanding how the topics came about. It is also fun in pretending to be time travelers going back with them in history in the rear mirror.

Every trades(wo)man carries a tool box full of tools. Combinations vary but a tape measure is there without exceptions. I started the book with a discussion of matter and non-matter. Matter is defined by dimensions and mass. In the matter physical world, dimensions are basic and a tape measure is indispensable in order to describe objects. Early physics was a study of motion and energy in the matter world. Motion involves time. The development of physics in historical terms was tied closely to contemporary technology at the time in both dimension and time keeping. Ancient scientists had neither analogue wristwatch nor digital time piece of today. They relied on slow celestial object movement and water clocks in Song Dynasty in China in 11th century to record events. After Archimedes discovered buoyancy in 200 BC, it waited until early 17th century when Christiann Huygens invented a self-contained escapement clock that was accurate and reliable enough for long distance marine navigation. Although the astrolabe (Samuel de Champlain) or sextant could help the ancient mariner locate his meridian of latitude by the sun, he still had no way to determine his meridian of longitude without an accurate clock to measure how far he had travelled from port after departure. Smallest segment of time he could measure was only 1 second mechanically. Precision was limited by available contemporary technology, as is still and always.

In Newtonian mechanics, an object in motion carries a kinetic energy of E = ½MV*2, where M is its mass, and V is its velocity at the time.

Albert Einstein was perhaps the most famous physicist of the 20th century. He was best known for his simple equation that every school child knows and reads out loud if asked,

E= mc*2, where c is the speed of light of 3 X 10*8 meters/ second, or 7½ times around Earth's equator/ second, and m (always **small**, never capitalized) is "equivalent" mass (= h/(cΛ) of his lesser known creation, the photon. Einstein created the "photon" as a phantom particle with a momentum (a quantum) that he used to explain his photoelectric effect in a billiard ball collision fashion. Clearly, Einstein was thinking of a radiation of wavelength (Λ) in the non-matter world that requires no acceleration for speed.

One may notice and ask immediately, what happens to the missing (½) that makes the two equations disagree ?

Professor A. Einstein's genius was in bridging between the two worlds of matter and non-matter. In matter world, object has mass that requires acceleration to pick up speed. Time spent for acceleration forces the use of

averaged velocity as the best possible in the equation for kinetic energy. That is where the (½) comes from in Newtonian physics of matter. In a non-matter world that has no mass and no acceleration required, everything happens **instantly**. Going back to the city water system analogy, water pressure is there already and it flows the moment a faucet is turned on. The word "instant" is unscientific to definite time, and speed of light (c) is the fastest so far scientifically established by available contemporary tools on hand. So fast that we call it "real time" when referencing to the clock. In the quantum-atomic world, particles and photons are interchangeable in Einstein's eyes. They move close to the speed of light, c (hence his theory of Relativity). That is too fast for Newtonian bigger objects and slower motion. Remember the elephant and squirrel race around the same track ? Both dimensions and time need a different set of tools. The tape measure and hour clock are too gross for accuracy. Real time there takes on a different meaning entirely. Tools on hand of contemporary technology in detecting location and speed of atomic particles incur errors comparable to data themselves. Hence the Heisenberg Uncertainty Principle that says that if one parameter, say location of the particle, is nailed down for maximum accuracy then it can be done only at the expense of equal maximum **in**accuracy in another associated parameter, in this case its velocity. If (Δd) and (Δp) were to represent margins of error for location and velocity of a particle being observed, then they bear a relationship of $(\Delta d) \times (\Delta p) = h$, Planck;s Constant to each other. Such lack of precision led some physicists to the use of waves to define position of a moving particle statistically. The wave shape represents the mathematical probability where the particle **may** be found, maximum at its centre of course but not with 100% certainty. By raising wave frequency, it narrows down wavelength for higher position accuracy. Blue-ray DVD presents sharper TV definition than helium-neon (HeNe) laser version due to its shorter wavelength, allowing finer dots and higher packing density on the disc by this same principle.

Professor A. Einstein was dissentient of Quantum Mechanics by his use of his phantom photon to explain photoelectric effect. His thoughts were classically Newtonian. Allegedly, he mocked or derided the Heisenberg Uncertainty Principle without professional courtesy. Some historians speculated that (Jewish-German) ethnic rivalry might possibly have entered into the fray. If it were true, then even world famous brilliant scientists might not be free of prejudice after all unfortunately.

Acknowledgement

Technically this book was a solo effort for the most part. That said, it could not have become reality without steadfast support and encouragement of my wife and partner Anne, both daughters Aisha, Emelie and our son Michael. If not materially, they were spiritually on board all the way in its evolution. I am most grateful to them.

I had the good fortune of having worked with many brilliant and generous colleagues who inspired, tolerated, trusted and challenged me on the most interesting projects that I could hope for. A complete list of their names would be endless and beyond my receding memory. Suffice it to single out those who figured prominently at the turning points in my life.

Dr. Foster, Director of Rutherford Cyclotron Laboratory of McGill University hired me during 1959 Christmas break to paint his corner office affording me an envied chance to see the legendary cyclotron up close and in person. Adrian van den Breckel, Maurice Bougie, Stan Rosenbaum, Bill Russell, Jeff May, Gordon McCallum (Northern Electric) took me in confidence on a leading edge R& D project in solid state material science. Dick Lindsay, Glen Swain, Geoff Avis (Microsystems International Ltd, MIL) and Wes Robinson, Vince Sheehan (MIL Penang, Malaysia) made my very first international assignment an unforgettable experience to be repeated many times in later years. Dr. David Kennedy, John Swift, Denis Colbourne, Mike Hollins (Bowmar Canada) gave me the opportunity to observe a Miqueladora (contract manufacturing) in Nogales, Arizona. I thank Dr. Raye E. Thomas and Darrel O'Shaughnessy for technology trips to India and Zimbabwe. Former Tanzania High Commissioner his Excellency Fadhil Mbaga and his wife Elizabeth were not only dear friends but gracious advisors as well during my many visits to their home nation for industrial development under CIDA sponsorship. Another dear friend and business-guru Meiz Majdoub of Accra, Ghana plus his nephew Eugene Asiedu truly opened my eyes to West African colourful history. I would be remiss without thanking Bill Kiss and Graham Neathway for those sparkling dialogues in summers 2003 and 2004 when prototyping and testing a 1 KW windmill generator. Retracting synapses began reconnecting. If not for so many clever friends in my life, I likely would have been a quiet science teacher in Hong Kong. I owe them so much for what I know now. Last but not least is my Giza Bedouin host who served me that memorable hot tea next to a 4000 year old pyramid in a hot African desert and in doing so inspired me to write this book half a century later. It is better late than never.

As a novice author, I accept all imperfections, glaring and subtle, in the book as mine. I shall do better if ever there should be a repeat attempt. Finally, your kind patronage is truly appreciated.

Select References

The list is in no particular order and by no means exhaustive. Yet, it is a foundation upon which my "AGBC Ionic Downburst" theory was framed and where my lines of thought were woven and threaded through.

1. Inorganic Chemistry, by J.R. Partington. Macmillan and Co. Ltd. London

2. Semiconductors and Semimetals, by Harold J. Hovel Academic Press

3. Physics of Semiconductor Devices, by S. M. Sze John Wiley & Sons

4. Physics of the atom, by Wehr-Richards Addison-Wesley Publishing

5. Elementary Plasma Physics, by Conrad L. Longmire John Wiley & Sons

6. Krakatoa, by Simon Winchester Harper Collins Publishers

7. Jamaica's Energy, by Raymond M Wright Petroleum Corp of Jamaica

8. Freemasonry & Birth of Modern Science, by Robert Lomas. Fair Wind Press

9. Map that changed the world, by Simon Winchester Harper Collins Publishers

10. The next 100 years, by George Friedman Griffin Press, Australia

11. 1421 and 1434, by Gavin Menzies Harper Collins Publishers

12. The Iron Rooster, by Paul Theroux Ivy Books, New York

13. Electricity & Magnetism, by Francis Weston Sears Addison-Wesley Publishing

14. Quantum Theory of Atomic Structure, by J. C. Slater McGraw Hill Book Co.

15. Waves, by C.A. Coulson Oliver and Boyd Ltd, Edinburgh

16. Quantum Mechanics, by Albert Messiah John Wiley & Sons

17. Quantum Mechanics, by Eugene Merzbacher John Wiley & Sons

18. Introduction to Q. M., by Chalmers and Sherwin Henry Holt & Company

19. Selected Problems in Q. M., by D. ter Haal John Wright & Sons Co., Bristol

20. Electricity & Magnetism, by M. Nelkon Edward Arnold Publisher, UK

21. Vacuum Deposition of Thin Films, by L. Holland Chapman and Hall Ltd., UK

22. Energy Storage, Electrochemical Society Proceedings 1976

23. Etching for Pattern Definition, Electrochemical Society Proceedings 1976

24. Principles of Vacuum Engineering, by M. Pirani + J. Yarwood. Chapman & Hall

25. Vacuum Technology, by Andrew Guthrie John Wiley & Sons

26. Introduc. to Electron Beam Technology, by Robert Bakish. John Wiley & Sons

27. Scanning Electron Microscope, by P. R. Thornton Chapman and Hall Ltd., UK

28. Electron & Ion Beam Science & Technology, Electroch Society Proceedings 1970

29. Heat, by M. Nelkon Blackie & Son Ltd.

30. Light and Sound, by M. Nelkon William Heinemann Ltd.

31. Elements of Physical Chemistry, by Samuel Gladstone Macmillan and Son Ltd.

32. Intermediate Chemistry, by T. M. Lowrey & A. C. Cavell Macmillan and Son Ltd

33. Mechanics, by Keith Symon Addison-Wesley Publishing

34. Introduc. to Geometrical & Physical Optics, by Joseph Morgan. The Maple Press

35. Radioactivity & Nuclear Physics, by James M, Cork. D. van Norstrand Co. Ltd.

36. Theory & Application of Microwaves, by A. B. Bronwell. Taiwan University

37. The Particles of Modern Physics, by J. D. Stranathan The Blakiston Company

38. Electronics, by Jacob Millman & Samuel Sewely Academic Press

39. Short Intermediate Mechanics, by D. Humphrey & J. Topping Longmans Green

40. Quick Calculus, by Daniel Klepper and Norman Ramsey John Wiley & Sons

The Author.

Macao born but raised in post-World War 2 Hong Kong, Nae Ismail came to McGill University and University of New Brunswick (UNB) to study molecular physics. He chose to stay after completion. His given name means "Godsend" in Arabic and was truncated for North American informal convenience. His Chinese name Ma Ka On in Cantonese or Ma Jah Aan in Mandarin/ Putonghua and his Muslim faith are traceable to his ancestry in Lanchow district of Gansu Province in West China.

Nae has earned graduate and post-graduate degrees in physics. His postgraduate research project was microwave spectroscopy in which angular momentum calculations were used to pinpoint locations of atoms in a molecule that is spun around its various rotational axes by pumping microwave energy into a waveguide in high vacuum and at low temperature. By superimposing an external electric or magnetic field to stretch the molecule into tiny perturbations, finer structural details are revealed.

After graduation and over next four decades, corporate assignments and consulting sent Nae to twenty eight countries in four continents fostering industrial partnership, offshore factory technology transfer and strategic planning. Apart from international exposure, his technical expertise includes the McGill cyclotron accelerator, 10 KW 270-degree e-beam processing, high temperature gas diffusion into elemental semiconductors and synthetic (GaAs, GaP, GaAsP) semimaterials for LED devices, thin film epitaxial layer deposition, plasmas sputtering, RF dry etching, high vacuum evaporation, YAG laser processing, HeNe laser interferometry, ultra-high vacuum, off-grid photovoltaics (PV), renewable solar energy and wind turbines. Nae ran his private solar firm for over two decades. He lives with his family in Ottawa, national capitol of Canada that is his adopted home.

Energy science is his lifelong passion. Other interests are pre-Renaissance Silk Road history, tectonic plate theory, astrophysics and marine explorations. This is his first book.

www.ingramcontent.com/pod-product-compliance
Lightning Source LLC
Chambersburg PA
CBHW032026170526
45157CB00002B/867